GROUNDWATER and the ENVIRONMENT

Applications for the Global Community

Igor S. Zektser

English Editor, Lorne G. Everett

Library of Congress Cataloging-in-Publication Data

Zektser, I.S. (Igor' Semenovich)
Groundwater and the environment / Igor S. Zektser; special editorial consultant Lorne G. Everett.
p. cm.
Includes bibliographical references and index.
ISBN 1-56670-383-2 (alk. paper)
1. Groundwater. 2. Water-supply. 3. Groundwater--Pollution. I. Everett, Lorne G. II. Title.

TD403.Z44 2000
333.91'04—dc21

00-030926
CIP

Direct all inquiries to CRC Press LLC, 2000 N.W. Corporate Blvd., Boca Raton, Florida 33431.

Lewis Publishers is an imprint of CRC Press LLC

No claim to original U.S. Government works
International Standard Book Number 1-56670-383-2
Library of Congress Card Number 00-030926
Printed in the United States of America 1 2 3 4 5 6 7 8 9 0
Printed on acid-free paper

Introduction

In recent decades, concern for the natural environment surrounding us not only has taken root in scientific literature, but also has appeared in our daily life. Many international symposia and conferences devoted to environmental protection have been organized by UN, UNESCO, UNEP and a number of international unions and associations. Specialists from many countries in the world express their concern in connection with increasing water and air contamination, land and wood resources depletion, and "rough" human interference in nature in general. In highly developed countries, there is, "less and less of surrounding nature, and more and more of the environment," as a poet said. This circumstance worries many people — not only scientists.

Different specialists — geographers, geologists, ecologists (environmentalists), soil scientists, hydraulic engineers, biologists, and economists — participate in solving the problem of environment protection and in elaborating on measures for the rational use of natural resources. In recent years, writers, journalists, and artists have taken part in discussing this problem. Such keen interest by the general public in "interrelations" between man and nature testifies that none of us is indifferent to the problem of the environment where we and our offspring will live.

Everyone should understand and remember that any type of anthropogenic activities from industrial, hydraulic and civil building, wood cutting, land improvement and plowing, surface and groundwater use, etc., to the realization of projects for regional water resources redistribution, always and inevitably affect the environment. The main purpose of science here is to correctly predict possible changes in the environment, to elaborate on the rational use of nature on a scientific basis, always being directed by the principle: "When using — protect, when protecting — use!" and to work out scientifically based recommendations for preventing a negative impact of man-induced activities on the environment.

In recent years, it has become evident in many countries that groundwater is one of our most important natural resources. Geologists used to say that groundwater is the number 1 natural resource. In 1931, well-known Russian academician A.P. Karpinsky wrote about groundwater, "Water is not only a mineral resource, not only a means for developing agriculture. Water is a real culture bearer; it is a living blood that creates life where there was none."

As opposed to other mineral resources, groundwater is characterized by some specific peculiarities that must be taken into account when assessing its reserves and determining perspectives for its use in national economies. Groundwater's main distinction is its renewability during total moisture circulation, which radically differentiates groundwater from all other mineral resources. Besides, and this is very important, when groundwater is exploited, not only is it pumped out, but also, in many cases, its regeneration is halted, caused by intensifying groundwater recharge with surface water and water from other layers, and also due to lessening evaporation from the groundwater table. Another significant characteristic of groundwater is its mobility and close connection with the environment. Groundwater is, on one hand, in constant and close interrelation with water-enclosing rocks, and, on the other hand, it is connected with surface streams, reservoirs, seas, landscapes, and vegetation. It is this groundwater singularity — its connection with the environment — that will be considered in this book. Here, it should be noted that groundwater, amounting to only 0.27% of the volume of the whole hydrosphere, plays an important role in satisfying man's water demand. This precious mineral resource is a subject for comprehensive and detailed investigations and generalizations.

Groundwater, being a part of the environment, is in complex and varied "interrelations" with its other components. Here, it should be noted that groundwater, particularly its intensive exploitation, significantly affects the environment. Thus, intensive groundwater withdrawal causes land surface subsidence, promotes activation of karst suffusion processes, affects river water content, and causes land drainage. Groundwater level determines the character of vegetation, affects productivity of crops, and dictates the necessity of drainage measures under buildings. Annual and perennial groundwater level fluctuations can cause flooding of urban and agricultural territories, promote landslide development, etc.

On the other hand, groundwater is affected by other components of the environment, particularly those that are under intensive anthropogenic impact. Thus, spring floods intensify groundwater recharge in the river valleys, and hence increase their natural resources. At the same time, surface runoff regulation by water reservoirs results in a decrease of flood intensity and duration, and causes a change in the regime of aquifer recharge, which reduces groundwater resources and deteriorates acting well-field exploitation. Vice versa, extensive development of land improvement, including irrigation and drainage, causes groundwater resources to increase in the adjoining territories.

This monograph is devoted to one of the most important ecological problems — the impact on the environment of intensive groundwater pumping. The compelling reason for writing this book was that the author, who has spent more than a quarter of a century engaged in studying groundwater, has repeatedly come in contact with erroneous concepts held by many people — including highly educated specialists in other spheres of science — about this significant component of the earth's natural wealth. The author has attempted to generalize and make analyses of available experience in intensive groundwater exploitation in different countries, primarily in Russia and the USA, and to assess the impact of such exploitation on the environment. According to the author, the purpose of hydrogeologists, when solving water supply problems, consists of correctly proving the norm that water of a required quality can be pumped out of an aquifer during an estimated period without any loss to the environment (including the groundwater itself), or at least, in a way that will make this loss minimal by special measures.

Preface to the English Edition

This document provides a very comprehensive insight into anthropogenic, hydrologic, and hydrogeologic activities that affect groundwater and the environment. Each aspect of the cause and hydrologic response is supported with international case histories or references to particular locations and countries where the phenomena occurred and can be documented. This is the first monograph to openly include the catastrophic groundwater-impact sites within the former Soviet Union. Specific locations, actual numbers, and examples of groundwater problems in Russia are described extensively. The international audience will, for the first time, begin to understand the magnitude of the environmental concerns within Russia. The document in its entirety will form the basis for an excellent understanding of the effects of anthropogenic activity on groundwater and allow individuals who have not been trained in the quantitative aspects of hydrogeology to appreciate many of the problems associated with pollution and water-resource development.

The quality of the English editorial review varies within the document. The consistency in the use of the English language varies substantially from chapter to chapter in the original manuscript. As a result, some chapters read extremely well, while others may be a little convoluted, as they are based on the Russian technique of sentence construction wherein very long, continuing sentences are the norm. Since a number of approaches and concepts did not lend themselves to a direct word-for-word translation, some of the sentence structure may require thought to fully appreciate the depth of the information to be found within specific sentences.

The monograph is a major contribution to our understanding of anthropogenic and climate-induced impacts on groundwater around the world. Consulting hydrogeologists, engineers, chemists, geologists, biologists, health officials, and government environmental administrators who have often wondered about the impact of the Soviet empire on its natural resources will find this document fascinating to read — perhaps even shocking, due to the magnitude of some of the identified pollution problems.

Lorne G. Everett, Ph.D., D.Sc.

Author

Igor S. Zektser received his doctorate in 1975 in hydrogeology science from All Union Research Institute for Hydrogeology and Engineering Geology in the former Soviet Union. For the past 30 years, he has been a director of the Laboratory of Hydrogeology in the Water Problems Institute of the Russian Academy of Sciences.

Dr. Zektser's fields of specialization include quantitative assessment of groundwater resources; evaluation of the contribution of groundwater to total water resources and water balance; assessment of the interrelation between ground and surface water and groundwater discharge into rivers, lakes, and seas; study of the main regularities of groundwater formation and distribution in various natural as well as disturbed conditions; and assessment of human impact on groundwater and groundwater vulnerability. He has published 10 monographs and more than 200 papers. Under his guidance as full professor, ten postgraduate candidates successfully earned their doctoral degrees.

Dr. Zektser was one of the original authors of the groundwater flow map of central and eastern Europe as well as the world hydrogeological map, and his research has had international recognition. He also served as expert and scientific leader for several projects of the International Hydrological Program, UNESCO, on groundwater resources assessment.

Dr. Zektser is president of the Russian National Committee of the International Association of Hydrological Sciences, a member of the Russian Academy of Natural Sciences, the New York Academy of Sciences, and the American Institute of Hydrology.

In 1991, Dr. Zektser was invited by the Environmental Protection Agency (EPA) to theVadose Zone Monitoring Laboratory in the Institute for Crustal Studies at the University of California in Santa Barbara to serve as a visiting research professor. From 1997 to 1998, he worked as a Fulbright Scholar at the same laboratory at UCSB.

English Editor

Lorne G. Everett is the Chancellor of Lakehead University in Thunder Bay, Ontario, Canada. Dr. Everett is also director of the Vadose Zone Monitoring Laboratory at the University of California at Santa Barbara (Level VI, which is described by the university as "reserved for scholars of great distinction") and Chief Scientist and Senior Vice President for the IT Group, Santa Barbara.

Dr. Everett earned a Ph.D. in hydrology at the University of Arizona in Tucson and a doctor of science degree (*honoris causa*) from Lakehead University in Canada for distinguished achievement in hydrology. In 1997, he received the Ivan A. Johnston Award from the American Society for Testing and Materials (ASTM) for outstanding contributions to hydrogeology. He is a member of the Russian Academy of Natural Sciences, and, in 1999, received its Kapitsa Gold Medal, the highest award given by the Academy for original contributions to science.

Dr. Everett is an internationally recognized expert who has conducted extensive research on subsurface characterization and remediation. He is a member of the Board of Directors and Chairman of the ASTM Task Committee on groundwater and vadose zone monitoring (D18.21.02). He chaired the Remediation Session of the first USSR/USA Conference on Environmental Hydrogeology (Leningrad, 1990). Dr. Everett has received numerous awards, published more than 150 technical papers, developed 11 national ASTM vadose zone monitoring standards, and authored several books including *Groundwater Monitoring, Vadose Zone Monitoring for Hazardous Waste Sites,* and *Subsurface Migration of Hazardous Waste*. EPA endorsed the book *Groundwater Monitoring* as establishing "the state of the art used by industry today," and the World Health Organization recommends it for all developing countries. Dr. Everett is also the author of the best-selling *Handbook of Vadose Characterization and Monitoring*.

Dr. Everett is editor of the Ann Arbor Press series Professional Groundwater and Hazardous Waste Science. He is coeditor of *Remediation Management*, the journal for environmental restoration professionals, and coeditor of the *World Groundwater Map* ,published by UNESCO.

Dr. Everett has made presentations before Congress and participates on Blue Ribbon peer review panels for Department of Energy (DOE) installations including Hanford (Washington), Lawrence Livermore National Laboratory (California), Yucca Mountain (Nevada), Rocky Flats (Colorado), Idaho National Engineering Laboratory, Fernald (Ohio), Savannah River (Georgia), and Oak Ridge National Laboratory (Tennessee). He is a member of the UC/LLNL Petroleum Hydrocarbon Panel, the DOE/EPA VOC Expert Committee, the Interagency DNAPL Consortium Science Advisory Board, and a Scientific Advisor to the U.S. Navy's National Hydrocarbon Test Site Program.

Acknowledgments

In recent years, many interesting publications on separate aspects of assessing groundwater withdrawal impact on the environment have appeared. The works of V.S. Kovalevsky, J. Margat, E. Custodio, A.A. Zhorov, V.L. Zlobina, P.E. LaMoreaux, Yu. O. Zeegofer, and other specialists to whom I have referred while preparing the present monograph, should be mentioned first of all.

The works published by the author in Russian and foreign periodicals personally or jointly with Doctors L.S. Yazvin, B.V. Borevsky, R.G. Dzhamalov, M.V. Kochetkov, L.G. Everett, H. Loaiciga, S. Cullen, and A.P. Belousova were also widely used in the preparation of the manuscript. I wish to express my profound gratitude for their collaboration.

In the present monograph, Chapter 8, "Groundwater Use and Public Health," was written by Dr. L.I. Elpiner; paragraph 5.2, "The Impact of Anthropogenic Activities on Groundwater," by the author jointly with Dr. L.I. Yazvin; and paragraph 6.1, "The Impact of Human Activity on Groundwater Resources," by the author in conjunction with Dr. A.P. Belousova; and paragraph 9, "The Impact of Human Activity on Groundwater Resources" — Dr. E.S. Dzektser.

I would like to thank the leadership of the Fulbright Program (USA) Institute for Crustal Studies at the University of California, as well as ARCADIS Geraghty & Miller Inc., whose support has made it possible for this book to appear. I also am pleased to thank O.A Karimova, T.N. Krashina, Tim Robinson, L.P. Novoselova, and L.N. Yakushin for their help in its preparation.

While preparing this book for publishing, much work was undertaken by its English editor, well-known American scientist Dr. Lorne G. Everett, to whom I express my sincere gratitude.

I.S. Zektser

To the blessed memory of my dear parents –
my first teachers in life and science.

Contents

1

The Problem of Fresh Water

Over the course of millennia, the idea that potable water was a vitally needed but inexhaustible and endless substance evolved. However, with population growth and development of industry and agriculture, freshwater demand has been abruptly increased and now its shortage is felt in many areas. Deficiencies in fresh water have been observed on about 60% of the earth's dry land. In many areas of the world, the water factor has begun to restrain the development of industry and agriculture.

What is the reason for storing fresh water? In some areas, it is due to natural climatic conditions (hot weather, drought, rare precipitation, absence of large water sources), in others, it is due to intensive and often ineffective water use for industrial purposes and, it is particularly important to emphasize, by progressive contamination of groundwater resources by industrial and agricultural wastes.

Enormous volumes of water are needed for industry and agriculture. Thus, a person consumes approximately 300–400 l of water per day for drinking and domestic purposes, 100,000 l are used for producing 1 t of sugar, and 150,000 l in steel manufacturing. And, on average, 200 l of water is required for growing 1 kg of vegetables. Preliminary predictions indicate that total water demand will increase tremendously in the world this year. For public water supply it will double; for industry, triple; and for agriculture it will increase by 1.5 times.

Scientists, specialists, public figures, and journalists express serious anxiety in connection with freshwater shortage. "Fresh water is a natural wealth, previously widely spread in most countries of the world, that, in the near future, will become less and less." Consider American ecologists, the authors of Environment in the Year 2000. "A threatening crisis is drawing near in potable water supply!" warns the Freiburg Institute of Ecology in its investigation of the distressing situation with water in Germany. "Hundreds of Regional Problems With Water Turn into a National Crisis," states the United States News and World Report.

According to the World Health Organization (WHO), about 1.2 billion people suffer from potable-water shortage. Particular concern is induced by worsening potable-water sanitary conditions that negatively affect considerable population groups in both developing and capitalist countries. WHO consequently demands urgent and drastic measures for improving this situation. About 80% of all cases of illness in developing countries are related to using water that does not correspond to a sanitary standard. The world public has spoken about "water starvation" of the planet, and about the approaching "water crisis." Water of quality has become an item of export. For instance, Hong Kong imports water by special pipelines from China, and in dry years, water is transported by tankers there. In some European countries, projects are considering the purchase of fresh water.

The problem of water has become international. According to the International Hydrological Program adopted by UNESCO, scientists and specialists from more than a hundred countries have combined their efforts in studying water resources for their rational use and protection from contamination and depletion, and are creating workable theory and methodology for water resources management.

When speaking about water storage, it is timely to indicate that society has already realized the importance of the water factor. In many cases, the availability of and possibility of using water resources determine the disposal of productive forces. Academician V.I. Vernadsky, marking the exclusive role of water in people's lives, wrote "Natural water involves and creates the whole human life. There could hardly be another natural body that would determine social structure, way of life, and existence to such an extent."

American scientist C.W. Fetter begins his famous book Applied Hydrogeology, which has printed several editions and is becoming a reference book for specialists of many countries, with the words: "Water is the elixir of life; without it life is not possible."

By the end of this decade, world water demand will be almost half of the global water runoff per year. When this happens, 18 out of 21 cities in the world with populations of about 10 million people in each, will be forced to satisfy their water requirements by means of remote water sources or deep groundwater pumping. Experts warn that water will become a dominating world problem in the next century, and difficulties with water supply can even become a threat to social stability in the world.

Special information dispersed in 1997 by the United Nations Organization on the occasion of World Water Day, which was marked by celebration in many countries, emphasized that almost all the largest cities in the world will enter the 21st century faced with a water crisis. For some big cities, problems of water supply are urgent enough already. The example of Mexico is widely known: during the last 10 years, land surface abatement has reached 10.7 m due to groundwater pumping.

Similar phenomena are observed in Houston, Bangkok, Jakarta and other coastal cities, where land surface decline induced by intensive water abstraction has caused sea water intrusion and flooding of large territories. With the coming of global warming, most coastal towns and island states may face a serious problem — a combination of the decline of land mass and flooding with the rising sea level.

One of the main reasons for the water crisis is water-resources contamination. Thus, in the Colorado River, which supplies water to many towns and inhabited localities of dry regions in the United States and Mexico, the concentration of salts has been almost doubled over the past 70 years. In the Volga River basin, the largest water artery in the European part of Russia, surface and groundwater pollution has reached such a degree that the government of Russia had to institute a special federal program called the Volga Revival.

In developing countries, only 5% of the industrial and domestic wastes produced in towns is subject to treatment and purification. The majority of two million tons of human excrement produced daily, as well as all the toxic and dangerous by-products of industrial production, are disposed into rivers and aquifers, thus contaminating them.

It should be noted that governments, specialists, and public citizens of highly developed countries realize the danger of water-resources contamination and depletion and take essential measures for their protection. Considerable work in this direction is being done in the USA, Germany, Great Britain, France, Austria, and the Scandinavian countries. Effective techniques for wastewater treatment and reuse have been developed in Stockholm.

It is not possible to present a very complex and multilayered water resources problem in its entirety, so the above data is given only to make the important conclusion that one of the main complex problems confronting scientists about the earth — primarily hydrology, hydrogeology and ecology — is to develop scientific bases for the rational use of water resources and their protection, both now and in the future. An essential place in the solution of this problem belongs to fresh groundwater, which, as one of the main components of the environment, is a reliable source for the supply of potable water in many regions.

2

Groundwater Function in Water Supplying Population, Industry and Agriculture

2.1 The Main Concepts

Groundwater occurs mainly in the rock masses of earth's upper crust. Depending on the character of the enclosing rock voids, groundwater is subdivided into:

- pore water — in sands, coarse gravel, and other sedimentary and detrital rocks
- fissured — in hard rocks and tightly cemented rocks (granite, sandstones), broken by fissures
- karst — in soluble rocks (limestones, dolomites, gypsum, etc.)

Rock layers saturated with water form aquifers. Several stepped and closely interrelated aquifers are called an *aquifer system*. Relatively impermeable layers, clays, massive loams, and non-fissured cemented rocks are identified as confining layers.

Here, it is proper to stop at the concepts and terms used in studying and exploiting groundwater in Russia, countries of the former USSR, and some countries in central and eastern Europe. These concepts, given in special documents and requirements for assessing mineral resources reserves, determine to a great extent the principles of assessing reserves and perspectives for fresh groundwater use.

There are some classifications for denoting groundwater volume, and in most of them the conceptions of "reserves" and "resources" are distinguished separately. The term "resource" was introduced into hydrogeology by academician F.P. Savarensky. The necessity of this concept for groundwater he proved by the fact that groundwater does not maintain a constant reserve like other mineral resources, as it is being constantly renewed in the process of common water circulation. When using groundwater, not only should the volume occupied by groundwater in this or that basin or water-bearing layer be taken into account, but also groundwater inflow (recharge) must be considered. For this reason, Savarensky considered it more correct to speak about groundwater "resources" and not "reserves," recognizing by this term groundwater inflow (recharge) and outflow (discharge), and using the term "reserves" to reflect only the volume of water in the basin or water-bearing layer that does not depend on its storage (capacity). Aquifer capacity (storage) and groundwater reserves in it need not be large, but productivity of the aquifer can be considerable if it is being recharged. And, vice versa, the groundwater basin can be large, but annual water inflow and its balance can appear small.

One more groundwater peculiarity should be noted here, connected with assessing the perspectives of its use. It lies in the fact that groundwater pumping depends not only on the volume of water in the layer and the flow into it under natural conditions, but also on the filtration properties of water-enclosing rocks, which determine resistance to groundwater flux to the well fields. Groundwater's peculiarities make it different from other mineral resources, and create the necessity of singling out some concepts of characterizing:

a. volume of water in the aquifer
b. volume of water flux into the aquifer under both natural and disturbed conditions
c. volume of water that can be abstracted by rational well fields for national economic needs

Thus, when assessing perspectives of exploitation for hard mineral resources or oil and gas it is sufficient to know their reserves, but that is not enough for assessing the possibilities for groundwater's efficient use.

Usually, natural groundwater reserves (static, age-old, geologic or storage are synonyms) characterizing the total volume of water in the layer in volumetric units are singled out. When assessing groundwater reserves in confined aquifers, "elastic reserves" are singled out, i.e., the volume of water released under drilling in a water-bearing layer, and reservoir pressure decrease in it under pumping or flowing due to water volumetric expansion and pore space, decrease. These reserves manifest themselves from the moment of drilling in the water-bearing layer up to the stabilization of the cone of depression and transition to a stationary regime of exploitation.

In the practice of hydrogeological investigations, groundwater natural resources and safe yield are assessed. Natural resources (dynamic resources is a synonym) characterize the volume of groundwater recharge due to infiltrating atmospheric precipitation, river runoff percolation and flux-out of other aquifers, which are totally expressed by the volume of flux discharge. Natural resources can also be manifested in the form of a water layer intersecting the groundwater layer. Thus, natural resources are an indicator of groundwater recharge, characterizing its main peculiarity as a renewable mineral resource.

Groundwater safe yield (resources) denotes a volume of water that can be pumped out of the aquifer per unit time by a technically and economically rational well field under a specified regime of exploitation and with water quality corresponding to requirements during the whole calculated period of exploitation. Thus, groundwater safe yield (resources) is one of the main criteria, determining possibility and expediency of groundwater use for different purposes. When making regional assessments, the term "potential reserves" is traditionally used, and when assessing water supply of particular objects, a "safe yield" is in use. When assessing safe yield (resources), a possibility is considered for using natural resources, including elastic reserves, and also involved (additional) resources being formed under well-field exploitation (involving surface water, groundwater of non-productive aquifers, etc.). Artificial resources, created by means of surface water submerging into natural subterranean reservoirs, using special installations, or being formed in zones affected by water reservoirs, in the irrigated territories, along the channels, etc. as a result of additional groundwater recharge from surface water sources, can serve as an important resource for a safe yield formation.

A relationship between different generic components of groundwater potential reserves becomes clear from the following general equation of groundwater balance in the exploited well field:

$$Q_y = Q_u + \frac{W}{\Delta t} + \Delta Q$$

where Q_y is a yield of exploited well field; Q_u is natural groundwater resources; W is water storage in a water-bearing layer, decreased under exploitation (i.e., natural-resources depletion, namely, layer drying within the cone of depression if the flux is unconfined, or elastic resources depletion if the flux is confined); Δt is calculated period of a well field exploitation; ΔQ is total additional resources, involved while exploiting.

If a cone of depression is stabilized or the exploitation period is unlimited ($\Delta t \rightarrow \infty$), then the second term of the above equation approaches zero. In this case, a well-field yield is caused by groundwater discharge, supplied by recharge and additional water flux ΔQ (if there are proper conditions for it).

In the first period of a well-field operation, potential reserves will be larger than natural ones due to depletion of natural groundwater resources, including storage and elastic ones. Under an unlimited exploitation period ($\Delta t \rightarrow \infty$) potential reserves will near natural resources by volume (under $\Delta Q = 0$).

Thus, groundwater natural resources are actually the upper limit that determines the recharge of constantly operating well fields with an unlimited exploitation period (excluding well fields where a yield is formed by additional reserves involved during exploitation).

It should be remembered that under exploitation, water-balance rearrangement occurs, caused by changing generic components of potential reserves. Development of a depression in the water-bearing layer under pumping can cause a water flux from the rivers, decrease evaporation from the groundwater surface, and cause or increase water seepage from the upper to the lower aquifers. Therefore, the function of groundwater natural resources as one of the generic components of potential reserves is different at different stages of the well-field operation.

Fresh groundwater resources are studied in two main directions:

1. Groundwater safe yield is explored and assessed for water supply of particular objects (town, plants).
2. Regional assessment of natural resources and potential reserves is made for perspective planning of possible groundwater use.

2.2 Current Status and Main Principles of Rational Groundwater Use

At present, fresh groundwater plays a substantial role in public water supply in many countries. An increase in the share of groundwater to the public water supply is observed. This is because groundwater has advantages over surface water. Groundwater generally contains micro- and macrocomponents needed for the human body; does not require expensive treatment; and is much better protected from contamination (including radioactive contamination), which is the most important factor.

Groundwater resources are much less susceptible to seasonal and long-term variations due to their regulated storage and because they are usually located near water users. Groundwater is essentially the only source of water supply in regions where surface water freezes or dries up. It is also important that well fields may be gradually put into operation as water demand increases, while the construction of hydraulic structures for surface-water development commonly requires overloading one-time investments. These circumstances and, in particular, a lesser vulnerability to contamination, predetermine a wider use of groundwater for public water supply.

This factor defined a substantial increase in groundwater use for public water supply in many countries in the second half of the 20th century. At present, groundwater is one of the main water sources in many European countries: in Austria, Belgium, Hungary, Germany, Denmark, Rumania, Switzerland, and the former Yugoslavia groundwater accounts for more than 70% of the total public water supply, and in Bulgaria, Italy, the Netherlands, Portugal, France, the Czech Republic, and Slovakia groundwater contribution ranges from 50 to 70%. In the USA, groundwater is the source for 75% of municipal water supply systems that provide the drinking water for more than a half of the country's population. In the last 30 years alone, groundwater use in municipal economies has more than tripled. Groundwater plays a great role in the water supply of China, Yemen, Saudi Arabia, Tunisia, Libya, and some other countries of Asia and Africa. The water supply of most cities and towns in Lithuania, Latvia, Estonia, Ukraine, Belorussia, Tadzhikistan, Armenia, and Georgia is based on groundwater.

During the last 25–30 years more than 300 million wells have been bored for water withdrawal in the world. In the USA, about a million wells are drilled annually, the water being used for domestic and public supply, irrigation, and technical needs. The depth of operating wells for water varies considerably and is determined by concrete hydrogeological conditions. Usually it is derived from 100 to 200 m, occasionally reaching 800–1000 m and even 2000 m in some cases.

The function of groundwater in supplying the water for towns in different countries and at different time periods has changed considerably. On the whole, in the initial stages of centralized water supply, spring (where it was possible) and river water were used as a source. Later, with increasing water demand, surface water was used more and more. Its progressive contamination in the second half of the 19th century and the resulting serious illnesses in the population, has made it necessary to reconstruct water supply systems in two ways: 1) improving the quality of water treatment, or 2) either complete or partial transition to underground water supply sources, including water of remote springs. The water supply system for Paris can be taken as an example. From 1865–1900, springs on the hill slopes some 80–150 km from the city were used, and surface water was used for technical water supply (Shevelev, Orlov, 1987). Hamburg is another example, where, after cholera outbreaks in the last century, surface water supply from the Elba River was replaced by groundwater. However, in the 20th century, with growing demand and limited groundwater resources, surface water use was considerably increased for the water supply of large towns in some regions. But its progressive contamination, and also cases of unforeseen emergency disposal of contaminants have made protected groundwater use timely. This tendency is now dominant in the strategy of organizing domestic and potable water supply.

The Moscow region is an obvious example of increasing groundwater withdrawal (Figure 2.1). As is clearly seen from the graph, after 1985 a tendency appears to decrease groundwater withdrawal due to ecological reasons related primarily to preventing contamination and depletion and also with strengthening requirements for environmental protection.

According to the European Economic Commission data (EEC) (Economic Bull., 1982), groundwater is the main source of municipal potable and domestic water supply in most European countries. Water supply of such large European cities as Budapest, Vienna, Hamburg, Copenhagen, Munich, and Rome (population a million people or more) is completely or almost completely based on groundwater, and for Amsterdam, Brussels, and Lisbon groundwater meets more than half of the common water demand.

Table 2.1 presents data on ground- and surface-water use for water supply of some large international cities.

At the same time, groundwater use as a source for centralized water supply is limited to a certain extent. Thus, in many cases, groundwater supply of large towns and cities, amounting to hundreds of thousands and even millions of cubic meters per day is unreal

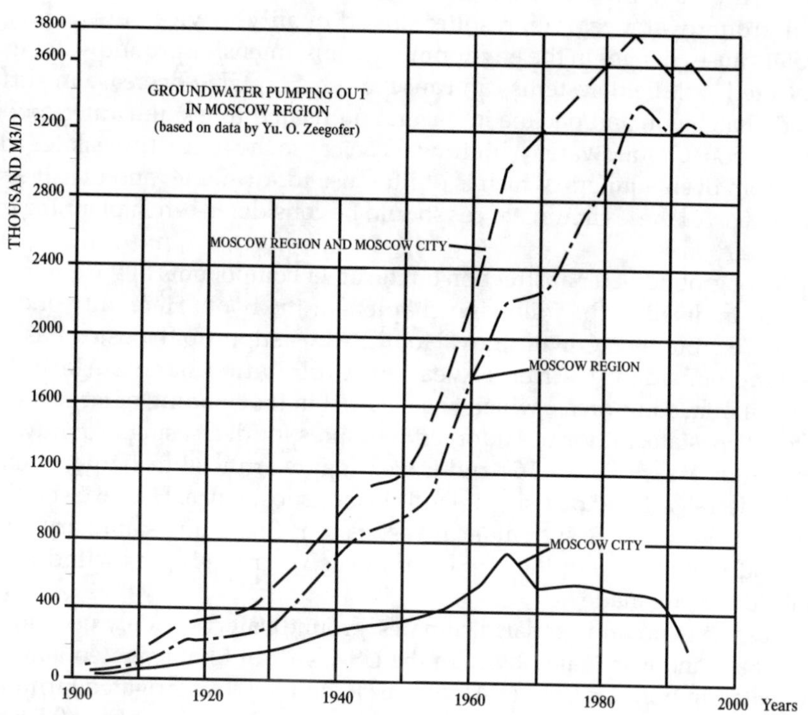

FIGURE 2.1
Groundwater pumping in the Moscow region, based on data by Yu. O. Zeegofer.

TABLE 2.1
Water Supply Sources of Large Towns in the World

Towns	Population	Sources	
		Surface water (%)	Groundwater (%)
Amsterdam	1.3	52	48
Antwerp	1.1	82	18
Barcelona	3.3	83	17
Berlin	5.6	58	42
Brussels	2.3	35	65
Vienna	1.7	5	95
Hamburg	3.6	-	95
Glasgow	5.2	63	37
Copenhagen	1.0	16	84
Libson	2.1	45	55
London	6.7	86	14
Madrid	4.1	91	9
Moscow	7.6	98	2
Munich	1.6	-	100
Paris	7.1	60	40
Rotterdam	1.4	90	10
Zurich	0.5	70	30
Tokyo	11.3	89	11
Chicago	5.9	88	12

because groundwater resources are limited, and because of the enormous cost of drilling hundreds and even thousands of water-pumping wells over a large area.

There is one more very significant aspect that should always be kept in mind when solving problems of groundwater use and that is closely connected with other environment components. Any changes in the amount of atmospheric precipitation inevitably results in changes in groundwater regime, resources, and quality — vice versa, changes in the groundwater cause changes in the environment. Thus, intensive groundwater exploitation by concentrated well field systems can cause an unacceptable decrease in surface water runoff, land abatement, and decline in vegetation related to groundwater and karst processes activization. Groundwater withdrawal can cause the influx (in-leakage) of mineralized water from deep aquifers, which is of little use for drinking, and of salt sea water in the coastal areas. All these circumstances should be considered when planning groundwater use.

The impact of groundwater abstraction on the other components of the environment will be considered in detail in the following chapters of the book. Here only one example is given. Intensive groundwater withdrawal for the water supply of Houston has caused considerable land subsidence. After 40 years of exploitation, land surface decline has amounted to 4 m in some areas, which has resulted in the flooding of large areas with sea water. After 1976, state authorities adopted measures for decreasing groundwater pumping in dangerous areas (Figure 2.2) and increasing its artificial recharge. Such measures substantially decreased the intensity of land surface subsidence. Here we should note that Texas is the only state where systematic investigations (including stationary observations, modeling, and predictions) of the impact of intensive groundwater withdrawal on land surface subsidence are made.

In countries with arid and semiarid climates, groundwater is widely used for irrigation. About 1/3 of all land is irrigated by it. In the USA, 45% of total irrigated land is irrigated by groundwater; in Iran it is 58%; in Algeria, 67%; and in Libya irrigated farming depends entirely on groundwater. In Russia, for land and pasture irrigation, about 0.5 km^3 of water is used per year. That is about 3% of the total groundwater withdrawal. Is that good or bad?

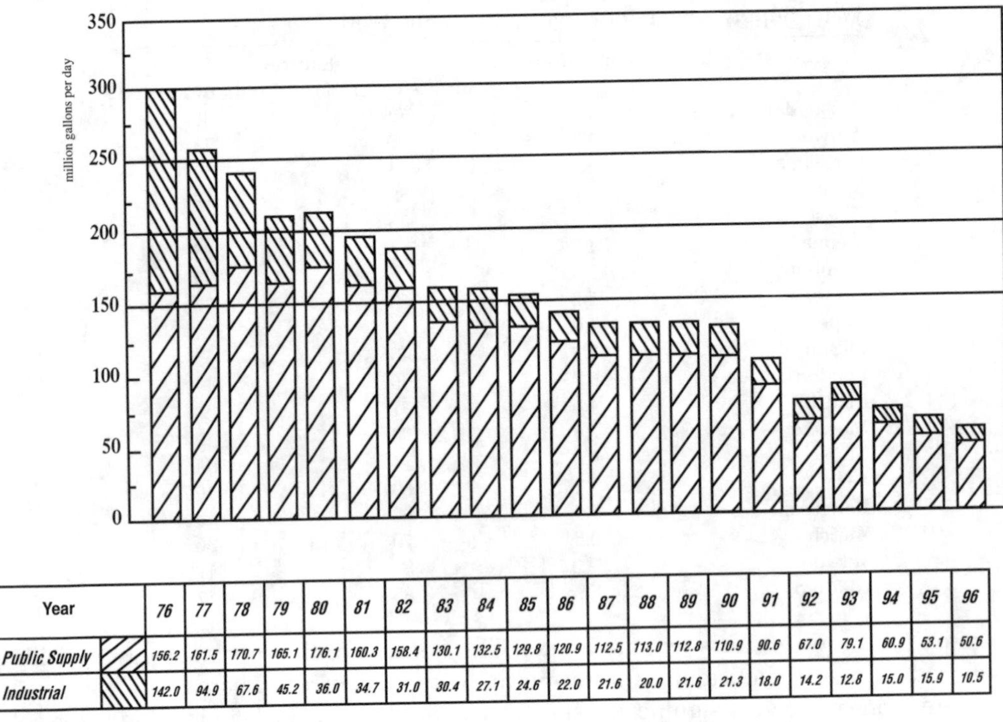

Year	76	77	78	79	80	81	82	83	84	85	86	87	88	89	90	91	92	93	94	95	96
Public Supply	156.2	161.5	170.7	165.1	176.1	160.3	158.4	130.1	132.5	129.8	120.9	112.5	113.0	112.8	110.9	90.6	67.0	79.1	60.9	53.1	50.6
Industrial	142.0	94.9	67.6	45.2	36.0	34.7	31.0	30.4	27.1	24.6	22.0	21.6	20.0	21.6	21.3	18.0	14.2	12.8	15.0	15.9	10.5

FIGURE 2.2
Groundwater pumping in the subsidence areas of Houston (from Groundwater Report, HGCSD, 1996).

In Russia (as in the former USSR) as well as in some other countries, there is strict water legislation. According to it, fresh groundwater of high quality must be used primarily for potable and domestic water supply. And, by special permission from nature-protecting organs, and only in the areas where assessed groundwater resources are sufficient to satisfy the present and most of the prospective potable water demand, is it possible to use fresh groundwater for other purposes (including irrigation) not connected with the potable water supply of the population. In essence, such legislation is true to our belief. We should aim at using surface water for non-potable water supply. Sometimes it may be more expensive than groundwater use, but it must be done primarily to preserve fresh groundwater of high quality for future generations. This principle, while shared by most specialists, is not always complied with. The state of California can be taken as an example. About 85% of the total groundwater resources is used for irrigation in this state. In the author's opinion, which is shared by most scientists — particularly hydrogeologists and ecologists — this situation is unacceptable from the viewpoint of rational groundwater use and it must be changed.

Finally, it should be emphasized that progressive surface water contamination has created a problem of increasing domestic and potable-water supply reliability. A wider use of fresh groundwater can be considered as one of the main means of improving the situation.

The use of groundwater safely protected from contamination is of strategic importance because it is the only source of water supply in extreme circumstances. This was demonstrated by the organization of the water supply in a number of towns and settlements in the zone influenced by the aftereffects of the accident at the Chernobyl Nuclear Power Plant.

In this connection, a possible complete or partial transfer of drinking-water supply systems to the use of safely protected groundwater sources is an important long-standing task for many regions.

When water demand cannot be fully met by safely protected water sources, a water supply strategy must envisage functioning of two sources including protected groundwater.

The possibility, advisability, and conditions for using groundwater as a source of public water supply are governed by a number of factors, the most important of which are as follows (Yazvin et al., 1994):

- availability of groundwater reservoirs whose safe yield makes it possible to fully or partially satisfy the requirements for drinking water and the possibility of the commercial development of these reservoirs
- natural quality of groundwater that may be used for public water supply directly or after a required treatment
- existence of natural protection of groundwater from surface anthropogenic contamination and conditions for creation of sanitary protection zones
- possible negative effects of groundwater withdrawal on other environmental components
- economic indices of the construction and operation of well fields including transportation costs

When planning groundwater use, different levels of aquifer protection should be considered. Groundwater in confined aquifers overlaid by uniform poorly permeable clay layers are almost fully protected from contaminants penetrating from the surface. In this case, contamination can result only from unsatisfactory technical conditions of well fields and test wells. Spring water in piedmont and mountainous areas is well protected from contamination if there are no anthropogenic activities in the recharge area. Groundwater from the surface aquifers is considerably less protected, particularly in the river valleys, where groundwater is closely related with surface water, and its exploitation causes the inflow of contaminated surface water. However, even in this situation, groundwater protection is much higher than the river water's, as contaminated water moving through the rock thickness is self-purified. Nevertheless, reliably protected confined groundwater in deep aquifers and spring water should be preferred in all cases.

A tendency for maximum groundwater use for potable-water supply of the population, particularly due to repeated unforeseen (emergency) cases of contamination of surface water sources must become the determinant in a general strategy for increasing reliability of domestic and potable water supply systems. This strategy must be aimed at the goal of reliably protected groundwater in water supply systems. Thus, in cases when settlements are supplied with surface water or insufficiently protected from groundwater contamination, measures should be adopted for a full change of water supply systems to reliably protected groundwater. When that is impossible, there should be an alternative source, based on reliably protected groundwater, that satisfies not less than 25–30% of potable water demand. Failing this, an autonomous potable water (and not domestic and potable) source should be created.

The most difficult problem of supplying people with ecologically pure potable water is organizing the water supply of large cities, with their enormous daily water demand. Full satisfaction of such demand using groundwater alone does not seem realistic in many cases, either because of limited natural resources or the huge cost of drilling wells over a vast territory. In such conditions, a strategy of water supply must be aimed at the functioning of two sources: surface and groundwater, and groundwater must meet an essential part of demand (not less than 25–30%). On the whole, three situations of groundwater use for satisfying domestic and potable water demand in larger cities can be singled out:

1. Domestic and potable water demand of large cities can be fully satisfied by reliably protected groundwater.
2. Large cities' demand for water can be partially satisfied by reliably protected groundwater.
3. Only an insufficient part of demand can be satisfied by reliably protected groundwater. In such cases, it is reasonable to investigate the possibility of creating autonomous sources of potable water supply based on the use of ecologically pure and protected-from-contamination groundwater.

Due to low usage of water by man for drinking and cooking purposes, ecologically pure and reliably protected groundwater can be found almost everywhere. It can serve as a base for creating an autonomous source of potable-water supply. The simplest way of organizing autonomous water supply is providing dwellings with potable water via pipes. However, for many reasons, primarily economic, this method can be realized only as a last resort. In this connection, building factories for bottling ecologically pure potable water with subsequent distribution should be considered the main source of autonomous potable water supply (Zektser, Yazvin, 1996).

Taken together, the irregular distribution of resources, different degrees of aquifer protection from contamination, availability of hydrogeological provinces with an increased content of separate normalized contamination and depletion of groundwater safe yield due to human activities, and possible negative impact of groundwater withdrawal on the other natural environment components predetermine a necessity for an individual approach to solving the problem of groundwater use in any particular case. The aim of hydrogeologists in solving water-supply problems is to properly substantiate the volume of water of a required quality that can be pumped out the aquifer during a calculated period without any loss to the environment (groundwater included) or making this loss minimum by special nature-protecting measures.

A combination of rational groundwater withdrawal, while meeting the quality standards that serve as a base for the concept of "rational groundwater use." It is recommended to understand its economically reasonable exploitation, providing groundwater safe yield protected from contamination and depletion and making it possible to preserve surface water and ecological conditions at an assigned level.

2.3 Groundwater Overexploitation

Due to increasing groundwater use and its withdrawal intensification, the problem of admissible limits for groundwater pumping of a aquifer to prevent its depletion becomes urgent. So, the term "groundwater depletion" should be defined. First, it is wrong to assume that any groundwater-level decline under pumping indicates its depletion. Groundwater use is actually impossible without a level (head) decline. But what limit of groundwater level decline will not cause irreversible changes in groundwater resources and quality?

To prove groundwater withdrawal possibilities and perspectives for water supply in Russia and countries of the former USSR, the concept of groundwater potential reserves has been put into the practice of hydrogeological prospecting and exploration. According to definition, *potential reserves* are a volume of groundwater that can be pumped during a calculated period (usually 25 years) by technically and economically rational well fields without a progressive level decline and water quality deterioration, and taking into account nature-protecting measures. The latter primarily concerns conservation of natural landscapes (rivers, soils, vegetation, etc.) to prevent or minimize negative withdrawal impacts on the environment. Assessment of potential reserves is based on special prospecting and exploration, test, and filtration works. Results of assessments are studied and tested by territorial and state Commissions for Mineral Resources Reserves. Only then is one allowed to drill wells and exploit groundwater for water supply and irrigation. It is assumed that aquifer depletion occurs only when a volume of potential reserves or a level decline resulting from exploitation exceeds maximum permissible value of decline.

The term "groundwater potential reserves" has, to our mind, disadvantages, primarily in the definition itself. What are "technically and economically rational well fields?" Possibly, what is considered "irrational" today — for instance, pumping out from larger depths — can appear rational tomorrow following advancement of drilling and pumping equipment. Therefore, the value of potential reserves as an indicator of the possibility to withdraw a certain groundwater volume, will change with time. That is not convenient when determining perspectives for groundwater use. Besides, in some cases, particularly when assessing the possibility of groundwater abstraction for water supply of a certain consumer, potential reserves are assessed only for satisfying a given concrete water demand, and thus do not characterize a general potential of a given aquifer or part of an aquifer. However, the concept of groundwater potential reserves is very significant and useful in practice, in spite of some disadvantages.

Close to it in the meaning is the concept of "balanced withdrawal" of groundwater, introduced by Hawaii Geological Survey, where fresh groundwater aquifers are underlain by sea water and where, consequently, it is particularly important to determine the volume of fresh groundwater that can be pumped without depletion of the aquifer and without water-quality deterioration (Sustainable development..., 1994). Here, "balanced "groundwater withdrawal is understood as an averaged volume of abstracted water not causing any damage to groundwater resources in the aquifer and not worsening groundwater quality or water point yield. This term is used for water pumping by wells, pits, infiltration galleries and other water-withdrawing constructions. For each aquifer, a balanced water withdrawal is determined by specifying equilibrium and time relations between recharge, natural discharge, water volume in rocks, and withdrawal. A balanced volume quantitatively is always less than a volume of recharge. As applied to Hawaii, the introduction of the concept of "balanced withdrawal" has become a governing principle for giving permission for groundwater withdrawal for different domestic purposes. When the main principles of a balanced water withdrawal are disturbed in Hawaii, there immediately emerges saltwater influx to the well fields, which causes an abrupt deterioration of abstracted freshwater quality. In some cases, when there is no direct connection between an aquifer and sea water, a full depletion of groundwater resources can occur.

In the last few years in some countries, primarily European ones with a delicate water balance, the problem of groundwater "overexploitation" exists.

The concept of "groundwater overexploitation" was put into use in the 1970s by renowned French hydrogeologist J. Margat. By "overexploitation," he meant groundwater withdrawal in volumes exceeding its recharge. However, he (Margat, 1982) later notes that the term "overexploitation" of an aquifer is contradictory and even ambiguous, as sometimes it is applied to a hydraulic concept of unbalanced exploitation that results in groundwater-resources depletion (withdrawal exceeds recharge), and sometimes to the multiaspect idea of overexploitation with undesirable consequences. It is not necessary to consider every withdrawal that exceeds recharge as overexploitation and to forbid it. There are many examples (to be discussed later) when, under a temporal water resources deficit and within the system frameworks, groundwater withdrawal exceeding recharge is consciously planned by the management. And vice versa, exploitation under dynamic equilibrium is not always free of undesirable consequences. E. Custodio (Custodio, 1982) is quite correct in noting that determination of overexploitation by comparing withdrawal and recharge is not always simple, and sometimes impractical due to indefiniteness of calculating recharge connected in particular with insufficient estimation of the time for transformation of infiltrating atmospheric precipitation into natural resources of an exploited aquifer, and also difficulties in estimating groundwater recharge changes caused by human activities (including artificial groundwater recharge).

It is reasonable to determine what should be understood as groundwater depletion. Russian specialists (e.g., L.S. Yazvin, B.V. Borevsky, N.I. Plotnikov, etc.) differentiate two concepts: "groundwater depletion" and "groundwater safe-yield depletion." Groundwater depletion occurs whenever water withdrawal exceeds its recharge and storage (gravitational and elastic) depletion occurs. A safe-yield depletion in the well fields occurs when the rate of groundwater level decline under exploitation exceeds maximum permissible rates specified when assessing safe yield. A principal difference between these two concepts is that under safe-yield depletion, measures must be taken to control its depletion (limiting groundwater withdrawal, its artificial recharge), and under groundwater depletion, these measures are not necessary and are needed only in individual cases. Moreover, in some social and economic conditions under extreme need, groundwater depletion can be planned analogous to a decrease of any other noncompensated mineral resources (Yazvin, Zektser, 1996c).

Safe-yield depletion may be caused by a withdrawal that exceeds the established yield and by changes in safe-yield generation conditions. Irrational groundwater development may also result in safe-yield depletion. For example, in the Crimea plain area, from the 1950s to the 1970s, a long-term, high-rate groundwater withdrawal (mainly for irrigation), which exceeded appreciably the safe yield, not only led to formation of deep cones of depression and a sharp regional decline of the water table, but also was responsible for salinization of water in the aquifers under development. This hydrogeological situation was somewhat improved by lowering a withdrawal rate and by artificial recharge in the 1980s.

A safe-yield depletion was also found in some well fields in the central part of the Moscow artesian basin. This may be explained by a change in ratios between groundwater and surface water involved in the development by induced recharge of well fields located at the river banks. Alluvial deposits may be clogged, resulting in a reduction of induced recharge and, therefore, in the lower productivity of the well fields.

Safe-yield depletion is often responsible for a change of groundwater quality. When the rate of withdrawal is greater than a determined safe yield, substandard water from other aquifers or surface water may encroach into the well fields. This is particularly important when the aquiferous strata are multilayered and the aquifers adjoining the aquifer under development contain mineralized water or water with an elevated concentration of a chemical component for which a standard has been established.

Protection of groundwater reserves from depletion includes regulating the volume of withdrawal according to estimated groundwater reserves; artificial recharge of reserves in operating well fields, including underground water reservoir construction; groundwater protection under hard mineral resources mining, including pumped out groundwater use for irrigation; technical and domestic potable-water supply; and regulation of groundwater-withdrawal assessment.

The available experience in considerable groundwater pumping indicates that if groundwater withdrawal exceeds its recharge, particularly if this excess is observed for relatively short time periods, it is nothing unusual and causes no serious troubles. Thus, the latest studies (Lopes Camago et al., 1991) have shown that in the Spanish peninsula and the Balearic Islands, 92 out of 369 groundwater basins where pumping exceeds recharge are considered overexploited. This excess is 650 million cu. m per year or 13% of the total pumped water in the area of 23 000 km^2. This situation is considered admissible and controlled. At present, only a few temporal declarations of overexploitation are accepted, and there is a definition of overexploitation in a new Spanish water law. Groundwater monitoring has been organized. Special attention is paid to the situation in the Balearic and Canary islands, where overexploitation can cause sea water entrapment (influx) in the well fields.

Numerous examples of overexploitation can be cited where groundwater levels were considerably declined as a result of groundwater pumping. The Ogallala aquifer in the USA, which occupies a territory of 135 000 km^2, has been overexploited for 60 years already. Here, groundwater levels in 20% of the territory have been reduced by 20–30 m, and groundwater withdrawal exceeds its recharge 10–15 times. Groundwater use for irrigation in this area is very profitable. At present, measures are being adopted for some limiting of groundwater withdrawal. In California, 42 out of 392 groundwater basins are considered overexploited. And though this circumstance does not particularly disturb American scientists, groundwater pumping was limited here from 5000 to 2500 million m^3 per year.

A clear example of essential groundwater level decline caused by "overexploitation" over several decades is the Moscow artesian basin.

In conclusion to this section it should be noted that, though the problem of groundwater overexploitation is not particularly disturbing from the viewpoint of groundwater-reserves depletion (if "overexploitation" is assumed to be the exceeding of groundwater withdrawal over recharge), nevertheless, it is given great attention, primarily when assessing the impact of considerable groundwater level decline on different components of the environment. It will be considered in detail in a following chapter. It should be noted that groundwater overexploitation was comprehensively discussed, using concrete examples, during the 23rd International Congress of the International Hydrogeological Association (Spain, April, 1991).

As an example of planned groundwater intensive withdrawal and nature-protecting measures, the problem of Moscow's water supply is considered. Until now, Moscow has been one of a few large cities in Russia whose water supply is totally based on surface water. Due to this, the population of the city is under a constant threat of malfunctioning water-supply infrastructure resulting from different emergency situations causing surface water contamination. The use of fresh groundwater protected from contamination in confined aquifers must substantially increase the reliability of domestic and potable water supply in Moscow.

Groundwater composition and properties have been studied within the Moscow region to the depth of about 1500 m. Fresh groundwater with mineralization of up to 1 g/l occurs at mean depths of 200–250 m — in some areas at only 80–100 m. Intensive groundwater use in this area was initiated in the second half of the last century. First, wells were drilled only into the first and second water-bearing zones from the surface aquifers in carboniferous deposits, and by the end of the century, they were drilled into all the main confined aquifers in the carboniferous zones. Due to the increasing groundwater withdrawal in the area, a special observation-well network has been drilled into confined and unconfined aquifers. At present, this network incorporates about 1100 observation wells.

Created in the mid-1980s permanent analogues are of great practical value, making it possible to constantly reproduce a hydrogeological situation within the Moscow region to assess the interaction of well fields, the impact of groundwater exploitation on the river runoff, and to calculate aquifers hydrogeological parameters (I.S. Pashkovsky, D.I. Efremov, Yu.O. Zeegofer). There are data from more than 1000 observation and operating wells in the model. Permanent analogues of the Moscow region make it possible to successively solve the problem of groundwater use management in this deficient — at least concerning water resources — region.

It should be noted that in Moscow and the Moscow region, which occupy jointly 0.3% of the Russian territory, about 12% of groundwater is withdrawn relative to the total volume of the groundwater pumped in the country (Zeegofer, 1995). Intensive groundwater withdrawal has resulted in the formation of huge and deep cones of depression.

At present, groundwater withdrawal within the Moscow region amounts to 4.0 million m^3 per day. In the last 3–4 decades, groundwater is practically unused for domestic and potable water supply, while its use in industry is considerable.

To solve the problem of a wider use of a good quality fresh groundwater, 4 large groundwater development areas have been explored at a distance of about 100–120 km from the city and a "General Scheme of a Combined Water Supply System for Moscow and Moscow Region using Groundwater Sources" has been elaborated. General groundwater withdrawal in the united system of water supply using four new large well fields is specified at a volume of 2.7 million m^3 per day. In this case, general groundwater withdrawal in the Moscow region must not exceed the amount of its natural resources (recharge), which are assessed at 8.7 million m^3 per day.

When distributing groundwater safe yield between Moscow and the Moscow region, priority is given to the Moscow region. Water demand for the area is 5 million m^3 per day. It is will be satisfied by groundwater in the development areas not included in the United System (3.8 million m^3 per day), and development areas included in the system (1.1 million m^3 per day). It is specified that only after satisfying a prospective demand in potable groundwater in the Moscow region is it possible to use it for Moscow (in the volume of about 1.6 million m^3 per day out of 2.7 million m^3 per day of the total withdrawal, specified by a joint system).

Current groundwater withdrawal is almost two times less than its exploitable (proved) reserves in the Moscow region. Here, in the newest well fields with groundwater exploitable (proved) reserves, water withdrawal amounts only to the first percentages of the proved (exploited) reserves, and in many well fields that have been operating for a long time, current withdrawal considerably exceeds proven (exploited) groundwater reserves. That results in groundwater-reserve depression, and in some cases, to its contamination.

When assessing groundwater safe yield, specified for use in the united water supply system of Moscow and the Moscow region, an interaction between operating well fields and explored sites was taken into account by calculations using mathematical modeling. Great attention in elaborating on the project of a united system was given to predicting the possible impact of so large a groundwater withdrawal on the environment. Numerous calculations, including modeling, indicated that, in separate sites, essential river runoff changes are possible, caused by intensive exploitation of groundwater. Concrete compensational measures are specified in the project (for instance, controlling dam construction in the Oka River) and also in some cases, decreasing of an earlier planned well-field yield or even a change of a site. It should be particularly noted that this project specifying groundwater withdrawal in large volumes has caused serious public response. Public opinion is worried about possible negative consequences of such a withdrawal, particularly the impact of groundwater withdrawal on the water content in small rivers, on vegetation and harvest, on karst-process activization, on forests, lakes, and ponds in the area.

Elaborating on "A General Scheme of United Water Supply System for Moscow and the Region, Using Groundwater Sources" has clearly indicated that there is no alternative to the wide groundwater use for Moscow and regional water supply, therefore Moscow and the region need a water supply source independently protected from contamination. The general scheme has received principal approval by scientists and specialists. However, accounting for a concerned public — inhabitants of towns in the Moscow region — decisions aimed at ecologically proving the perspectives of groundwater withdrawal and organization of monitoring for the main environmental components in the Moscow region have been accepted.

Only after obtaining reliable proven results will the problem of building concrete well fields for groundwater withdrawal be solved, with simultaneous realization of necessary measures for preventing or minimizing possible negative consequences of considerable groundwater withdrawal.

2.4 Groundwater Resources of Russia and Their Use

2.4.1 Regional Investigations: State of the Art

Regional investigations of groundwater resources will be briefly presented here.

In the former USSR, for the first time in the world, significant investigations have been carried out for regional assessment of natural and predicted groundwater potential reserves. They were initiated at the end of the 1950s following concrete practical demands to give a quantitative assessment of groundwater use perspectives for proving the "Scheme of Water Resources Complex Use and Protection in the Country." As a result, in a short period of time, regional quantitative assessment of fresh groundwater natural resources of the territory was made, concluding by publishing a monograph and a series of maps for groundwater discharge in the USSR. For the first time, values of natural resources have been obtained for the territory of the country and its large regions, the main regularities of its formation depending on physical–geographical and geological–hydrogeological conditions, have been revealed, and time and space peculiarities of specific values and coefficient changes characterizing groundwater discharge have been determined. The methods used for regional assessment of groundwater discharge and natural resources allows us to objectively and economically effectively assess them by analyzing the available hydrological and hydrogeological materials without special exploration works (Chapter 3).

In the published maps of groundwater discharge in the former USSR territory there are mean perennial groundwater-discharge modules characterizing groundwater discharge from the area of 1 km^2; groundwater discharge coefficients giving a ratio of groundwater discharge to atmospheric precipitation; and coefficients of river recharge with groundwater, giving a portion of groundwater in the total river runoff. Besides natural conditions and factors determining the regularities of natural-resources formation (composition and stratigraphy of water-bearing rocks, areas of karst development, spreading of freshwater lenses, areas of surface water absorption, etc.) are given in the maps.

Based on the principle of natural water unity, groundwater-discharge maps make it possible to solve the following practical problems connected with the complex use and protection of water resources:

- determining groundwater discharge and natural fresh groundwater resources for assessing water supply of separate areas in the country
- determining the value of groundwater discharge into rivers for characterizing the groundwater component of river runoff as the most stable surface water resources part, and also for predicting changes in river runoff and the ecological situation under intensive groundwater withdrawal
- determining the value of groundwater recharge for regional assessment of its potential reserves while compiling water management balances of economic regions and natural-territorial complexes
- determining the value of groundwater discharge as an element of water balance in the country and its separate regions under prospective-planning water-resources complex use and protection

Experience in regional assessment and mapping of groundwater discharge in the USSR has received international acknowledgment. Corresponding to the international hydrological program adopted by UNESCO, specialists from some European countries studied conditions of formation, quantitative estimation, and mapping groundwater discharge in the territory of central and eastern Europe. The result of this work is the Map of Groundwater Discharge in Central and Eastern Europe, at a scale of 1:1 500 000 (Groundwater Flow Map …, 1983) and the monograph to it (Groundwater Flow of Area …, 1982).

In the last few years, significant international work on regional assessing and mapping of natural groundwater resources around the globe has been completed. Specialists from many countries in the world participated in this work and as a result, in 1997, the Map of Hydrogeological Conditions and Groundwater Discharge in the World, at a scale of 1:10,000,000 was published in the USA (chief editors R.G. Dzhamalov and I.S. Zektser).

These investigations were a contribution to the International Hydrological Program, UNESCO, and can be an example of joint efforts of scientists from different countries in solving important scientific problems.

Regional estimation of groundwater safe yield is made by determining the volume of groundwater withdrawal from the aquifer, provided that groundwater level decline by the end of exploitation must not exceed a specified value (determined in advance, basing the data on water-bearing-layer parameters), and water quality must satisfy certain standards. Under regional estimation, calculation of both potential and predicted potential reserves is usually made. What is the difference between them? Potential exploited reserves characterize maximum possible groundwater withdrawal out of the aquifer, and predicted resources point at possible groundwater use under a certain location of consumers or considering a concrete water demand. Here, a regional assessment of predicted potential reserves is made either for a conditional well-field location, or (if it is known) taking into account a scheme of particular water consumers' location and water demand.

Lately, considerable work has been done in Russia on regional assessment of groundwater potential reserves of artesian basins. Maps at different scales with potential reserves moduli have been compiled. Module of potential reserves means water yield that can be obtained from 1 sq. km of the aquifer area. Under regional estimation of potential reserves for separate prospective regions, water demand of concrete water consumers and possible location of future well fields were taken into consideration. As a result, for most hydrogeological regions of the country, a principal possibility for groundwater use has been revealed and a base has been made for planning prospecting and exploration works for water supply of concrete objects. However, it should be noted that a decision for designing and drilling groundwater well fields is taken up, not based on the results of regional assessment of its natural or potential reserves but only after conducting special works with an obligatory approval of groundwater safe yield by a state or territorial commission for mineral resources. According to normative documents of the State Commission for Mineral Resources Reserves, groundwater safe yields are subdivided into categories according to a degree of their study. Here, categories A and B are singled out as those most studied, allowing the right to drill groundwater well fields. The amount and technique of prospecting and exploration work (including drilling the wells and their test operation, observation of groundwater level, and chemical composition changes, etc.) and also ways for calculating groundwater storage depend primarily on concrete hydrogeological conditions and study of the area for work.

Accumulating and generalizing the experience of groundwater exploration in different hydrogeological conditions and data analysis obtained by exploiting some large well fields, have created a necessity of developing some theoretical aspects of groundwater dynamics. It results in creating quite new bases for exploration and assessment of groundwater storage, based on the theory of transient movement (flow), elastic regime of filtration, and penetration through poorly permeable deposits. Developing a concept of the aquifer boundary conditions as factors determining regularities of groundwater storage formation and principles for making schematic hydrogeological conditions for calculating the storage were very important. While assessing groundwater storage, methods of mathematical modeling are widely used that provide for increasing their reliability and making prospecting–exploration and test-filtration works rational.

Investigations carried out in recent years under the guidance of L.S. Yazvin and B.V. Borevsky made it possible to prove the concept of groundwater development areas, to improve the technique of prospecting and exploration work, and to develop a more grounded safe-yield classification.

At present, the concept of groundwater development area is widely used in practice. A "groundwater development area" is the part of an area of aquifers or aquifer systems occurrence, within the limits of which, under the impact of natural and artificial factors, if compared with surrounding territory, most favorable conditions are created for groundwater withdrawal at a volume, enough for its reasonable use in the national economy.

In Russia, for domestic, potable, technical water supply and land irrigation, some thousands of fresh and saline groundwater development areas have been explored, and their safe yield approved in the State and Territorial Commissions for Mineral Resource Reserves.

2.4.2 The Main Regularities of Groundwater Natural Resources Formation and Distribution

As it has already been noted, in recent decades in some countries, significant investigations have been carried out for regional assessment and mapping of groundwater natural resources and discharge in some large regions. As a result of these investigations, maps of groundwater discharge at different scales have been compiled and published. Among them, the Map of Groundwater Discharge in the USSR Territory, at a scale of 1:2 500 000 (1978) should be mentioned first, as well as the Map of Groundwater Discharge in Central and Eastern Europe, at a scale of 1:1 500 000 (1982), the Map of Hydrogeological Conditions and Groundwater Discharge of the Land in the Globe at a scale of 1:10 000 000 (1997). For some artesian basins and hydrogeological structures, maps have been compiled at a large scale. Analysis of the available maps at different scales in the former USSR territory and some bordering makes it possible to determine the main regularities of forming and distributing natural groundwater resources (groundwater discharge) in various natural and climate geological and hydrogeological conditions (Kudelin, 1960; Zektser, 1977; Vsevolozhsky, 1983; Groundwater discharge in the USSR territory; Dzhamalov, 1973; Lebedeva, 1972; et al).

In short, these regularities can be characterized as follows. Distribution of the main quantitative groundwater discharge characteristics in the territory of regions is extremely inhomogeneous and is differentiated by the main geological-structural elements and landscape and climate zones. The most common regularity is a different character of distributing parameters of groundwater discharge within platform (plain) territories and mountainous-folded structures at a range of changes from less than 0.1 to 6.0–6.8 l/sec and from 0.1 to 30–50 l/sec•km^2 correspondingly. In the USSR territory, more than 55% of total groundwater discharge volume is formed within mountainous-folded areas, about 42% corresponds to vast platforms (Russian, West Siberian, Turanean) and only about 3–4% of total groundwater-discharge volume comes for crystal shields.

Analysis of spreading the volume of groundwater discharge in the main landscape and climatic zones shows that more than 80% in the total groundwater discharge volume is confined to excessively moistened and humid zones, about 18% of the discharge is formed in insufficiently moistened ones, and only about 2% in the arid zones.

In a continental platform, a regular spreading of groundwater-discharge parameters is characteristic according to the effect of latitudinal climatic factors. On the background of latitudinal distribution, local changes of groundwater discharge values, caused by hydrogeological structure of intensive water-exchange zone (cross section) and geofiltration environment types are most distinctly manifested. Thus, maximum values of groundwater-discharge modules are characteristic for intensive karst-development areas, and, to a lesser degree, for areas where the upper part of the cross section is constituted by coarse detrital or sandy fluvio-glacial and recessional moraine deposits, and also for river valleys, constituted by easily permeable alluvial deposits. Minimal module values of groundwater discharge have been registered for areas where a zone of intensive water exchange is constituted by loamy and clayey rocks, and for relatively lowered, weakly dissected territories where poorly permeable rocks in the upper part make the infiltration of atmospheric precipitation difficult.

In folded-mountainous areas, the distribution of groundwater-discharge values is mainly caused by sharp changes of medium geofiltration types and orographic precipitation increase with the locality elevation. Thus, high groundwater discharge modules in the Caucasus, Carpathian and Balkans are caused by widespread permeable fissured rocks in folded-mountainous deposits and highly permeable coarse detrital deposits in the intermontane depressions that in complex with a deep erosional dissection of the relief and

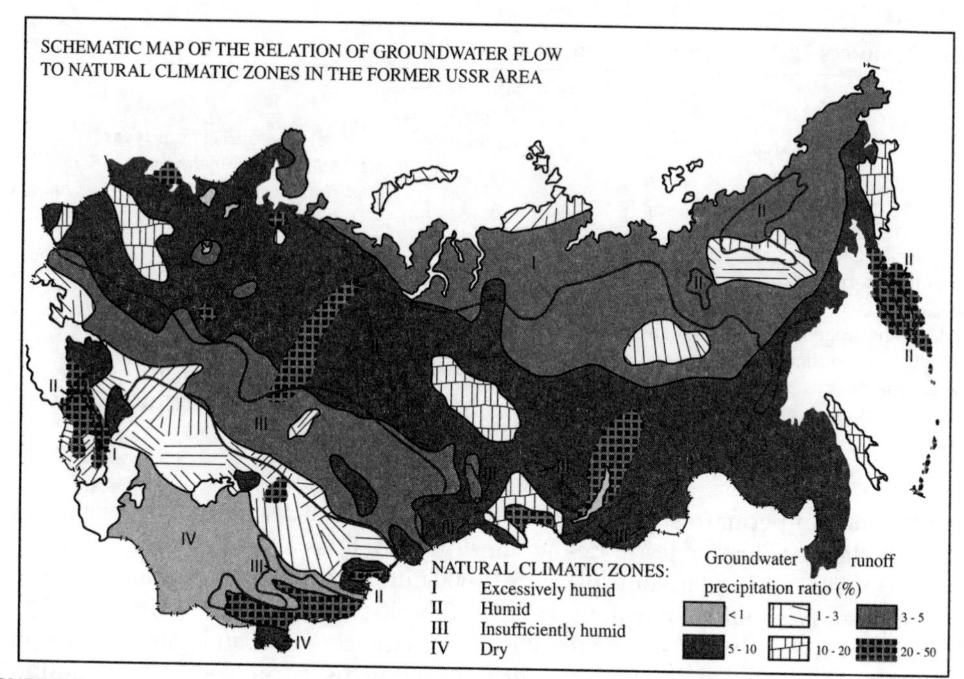

FIGURE 2.3
Schematic map of the relation of groundwater flow to natural climatic zones in the former USSR area.

essential volumes of atmospheric precipitation ensure favorable conditions for groundwater recharge.

The groundwater function in the formation of a water balance and water resources in regions is quantitatively characterized by groundwater-discharge coefficients and coefficients of groundwater influx into the rivers.

A groundwater-discharge coefficient is a ratio between groundwater discharge and atmospheric precipitation. It indicates what part of atmospheric precipitation (usually in percentages) contributes to groundwater recharge. This coefficient within a country is, on the average, 9% and ranges from 1% to 50% and more. The main peculiarities of groundwater discharge coefficients spreading are caused by the impact of a complex of natural factors, a ratio between atmospheric precipitation and evaporation, structure and thickness of rocks in the vadose zone being of prime importance. For a plain territory, latitudinal zoning is observed as a general regularity; that is, groundwater discharge decreasing from the northwest and to the southeast from 10–12% in excessively humid zones to 1% and less in steppe and semi-arid zones (Figure 2.3).

Separate areas with raised groundwater-discharge coefficients are observed on this background. First it is observed in the highlands, due to the abundance of atmospheric precipitation here and improvement of infiltration conditions (Valdai, Sredne-Russkaya, Privolzhskaya uplands, Enisei range, Severo-Baikal highlands, etc.). In mountainous areas, simultaneous with the increasing amount of atmospheric precipitation and locality height (up to certain limits) groundwater discharge coefficients also increase. Thus, in the Carpathians, they increase from 5% to 10–15%, in the Urals from 10% to 20–40%, in the Altai from 5–10% to 15–20%. In the Caucasus and in the Middle Asian mountains, the increase of groundwater discharge coefficients is most substantial (up to 25–35%).

In areas of karst development, groundwater discharge coefficients are also high (to 30–40% and more in the Siluriisk plateau, in the Onega-Severodvinsk interfluve, Kuloisk plateau, Timan).

TABLE 2.2

Water Resources Distribution in the Largest Areas

Area	Square, thous. sq. km	Resources, km³/year		
		Surface water (total river runoff)	*Groundwater (groundwater discharge)*	*Ratio between ground and surface water*
Russian Platform	4060	719	170.0	0.24
The Caucasus	264	92	45.7	0.50
The Urals hydrogeological folded area	462	79	25.4	0.32
West Siberian artesian area	2919	583	136.9	0.23
Siberian platform	3500	738	108.9	0.15
Central Kazakhstan hydrogeological folded area	938	67	65.0	0.16
Verkhoyansk-Chukostsk hydrogeological area	2420	538	65.0	0.12

The function of permafrost in spreading the groundwater-discharge coefficients is very distinct. In the vast Siberian territories and northwest of Russia, where atmospheric precipitation is 300–400 mm and in some places 500–600 mm, groundwater-discharge coefficients are not big — 5%, and then only in the south of Siberia (Severnoe Pribaikalie and offshoots of Verkhoyansk ridge); where permafrost is of an island character and annual atmospheric precipitation increases up to 800 mm, groundwater-discharge coefficients amount to 15–20%.

Coefficients of groundwater recharge of rivers, which indicate groundwater portion in the total river runoff, are also an important characteristic, and make it possible to determine a relation between ground and surface water resources in many areas of a humid zone. These values are given in Table 2.2 for the largest areas.

On the average, a groundwater discharge coefficient is 24%, changing from 5–10% in areas with relatively thin zones of intensive water exchange, weakly dissected relief and favorable conditions for surface runoff formation up to 40–50% and more in areas constituted by rocks with a high water content and intensively drained by rivers.

The analysis of relations between groundwater discharge and common river runoff has become of great practical value in the problem of complex water-resources use, in particular when making water management balances for separate regions and assessing the impact of groundwater exploitation on river runoff.

2.4.3 Perspectives for Groundwater Use

Let us consider in detail the present-day situation with groundwater resources in Russia and their use.

Russia possesses considerable fresh groundwater potential reserves, their total amount exceeding 230 km³ per year. Groundwater safe yield of the explored development areas amounts to 27 m³ per year, 18.4 km³ per year of which are prepared for industrial exploitation. However, groundwater withdrawal is almost twice as low as reserves (storage) prepared for use, which indicates considerable perspectives in increasing groundwater withdrawal for different purposes (Yazvin et al., 1996).

At present, the necessity to intensify groundwater use for increasing the reliability of systems for domestic and potable water supply of the population in Russia is generally acknowledged. In a special federal program called "Providing the Population of Russia with Potable Water" a considerable increase of groundwater withdrawal and use is specified.

TABLE 2.3

Groundwater Withdrawal and Use, km³/year, in Rukssia (Yazvin, Zekster, 1996)

Economic Region	Withdrawn Water			Groundwater use			
	total	mine and quarry outflow	discharge without use*	total	public water supply	technical water supply	irrigation and pasture watering
Northern	0.36	0.17	0.19	0.17	0.12	0.05	-
Northwestern	0.39	0.08	0.10	0.29	0.18	0.11	-
Central	3.41	0.29	0.32	3.09	2.37	0.72	-
Volga-Vyatka	0.50	-	0.01	0.49	0.34	0.15	-
Central Chernozem	1.43	0.18	0.12	1.31	1.14	0.15	0.02
Volga	1.37	0.02	0.01	1.36	0.84	0.40	0.12
North Caucasian	1.95	0.06	0.18	1.77	1.52	0.21	0.04
Ural	2.58	0.64	0.55	2.03	1.66	0.37	0.01
West Siberian	1.93	0.38	0.32	1.61	1.04	0.36	0.21
East Siberian	1.22	0.21	0.21	1.01	0.65	0.22	0.14
Far Eastern	0.74	0.11	0.12	0.62	0.44	0.18	-
Kaliningrad' Region	0.10	0.02	0.08	0.06	0.02	-	
Total for Russia	15.98	2.16	2.15	13.83	10.36	2.93	0.54

*This discharge includes water losses (e.g., in transportation), water outflow from flowing wells and vertical drainage wells, and outflow from mines and quarries.

The analysis of groundwater use in the former USSR territory results in a conclusion that in the latest decades its withdrawal for public use has essentially increased. In the mid-1950s, an increase in the use of groundwater occurred. If at the beginning of the 1950s, groundwater withdrawal for this purpose was about 10–15% of the total consumption of surface and groundwater, and by the end of the 1960s, it increases to 44%, and at present it is about 53%. In the period from 1980 to 1989, groundwater use for public, agricultural, and industrial water supply increased from 8.8 to 13 km³ per year. A tendency to increasing groundwater withdrawal for public water supply will remain in the future due to a necessity of supplying the population with fresh groundwater.

Table 2.2 shows that public (municipal and rural) water supply is the largest groundwater consumer; it amounts to approximately 10.4 km³ per year (about 75% of total groundwater use). Groundwater accounts for approximately 57% of the total municipal and rural water supply. However, the municipal water supply proper uses no more than 4.9 km³ per year of groundwater, making up 35% of the total withdrawal of surface and groundwater and amounting to 14.4 km³ per year. Another 2.9 km³ per year is used for municipal needs by well fields belonging to departments managing individual industries (Yazvin and Zektser, 1996).

Groundwater is being developed rather irregularly over the area of the Russian Federation. It is being intensively used in the Bryansk, Vladimir, Tver', Kaluga, Oryol, Smolensk, Tula, Belgorod, Voronezh, Kursk, Lipetsk, Tambov, Orenburg, Tomsk, Tyumen', Chita, Amur, Kamchatka, and Sakhalin regions; in Krasnodar, Altai, and Krasnoyarsk territories; and in the Marii-El, Mordoviya, Kabardino-Balkariya, Chechnya, Ingushetiya, Northern Osetiya, Mountain Altai, Bashkortostan, and Buryatiya republics. In all these regions, groundwater use ranges from 70 to 100% of the total water use. Groundwater is poorly used in Novgorod, Arkhangel'sk, Murmansk, Kostroma, Yaroslavl', Astrakhan', and Omsk regions, and in the Karelian Republic, where its use accounts for 3–20%.

At the present time, more than 60% of cities and towns in the Russian Federation have groundwater sources of water supply (Table 2.3). Groundwater is the main source of water supply in small and average towns, however, in some regions, its function in domestic and potable water supply of large cities, including those with populations exceeding 1 million

TABLE 2.4

Water Supply of Cities and Towns, % with Different Population (Yazvin, Zektser, 1996)

Water-supply sources	Less than 50ths	51-100ths	101-250ths	251-500ths	500ths-1mln	Over 1mlm
Mainly groundwater (more than 90%)	74	57	46	37	3	0
Mainly surface water (over 90%)	15	21	24	37	39	82
Ground and surface water	11	22	30	26	28	18

people, is very significant. Still, some large cities of the former Soviet Union, including Moscow, St. Petersburg, Samara, Perm, Ekaterinburg, Zaparozhye, Omsk, Dnepropetrovsk, Odessa, Yaroslavl, Rostov-na-Donu actually use no or few sources of groundwater supply and are under a constant threat of possible quality deterioration of the water consumed. Some towns, where water supply is based mainly on groundwater, but exploited aquifers are closely related with surface water courses and are actually not protected from contamination, also belong to this group

Let us discuss in more detail a pattern of public water supply in the largest cities of Russia (with a population of more than 250 000) as noted in Table 2.4. Water supply of 34 from 77 cities is predominantly based on surface water (more than 90%), and 24 cities meet their water demands mainly by groundwater (more than 90%). These cities are Bryansk, Tver', Oryol, Smolensk, Tula, Dzerzhinsk, Saransk, Voronezh, Kursk, Belgorod, Lipetsk, Tambov, Tol'yatti, Krasnodar, Sochi, Groznyi, Vladikavkaz, Orsk, Magnitogorsk, Novokuznetsk, Tomsk, Krasnoyarsk, Chita, and Ulan-Ude. The remaining 19 cities have combined water-supply sources.

Groundwater contribution to water supply accounts for 79–85% in Kaluga, Nal'chik, Orenburg, Ufa, and Komsomol'sk; 39–53% in Ivanovo, Ryazan', Taganrog, Barnaul, and Tyumen'; and 15–28% in Vladimir, Rybinsk, Penza, Ul'yanovsk, Kazan, Kemerovo, Bratsk, Khabarovsk, and Kaliningrad.

The irregular use of fresh groundwater in the Russian Federation is, above all, due to the differences in hydrological and hydrogeological settings that govern safe groundwater yield and groundwater recharge, under both natural and disturbed conditions.

It is important to note that the increase in groundwater use in the recent years averaged 0.2 km^3 per year. This amount is a third of that used in the early 1980s. This is mainly due to a general decline in production and insufficient funding of water-management needs. However, in some regions, groundwater development is restrained by an incorrect strategy of water-management organizations, which prefer to use surface water for public water supply even in regions with groundwater resources that are protected from contamination and ready for development.

For comparison, it should be indicated that groundwater met approximately 60% of all public water demands in the area of the former USSR in the late 1980s. Water supply of the majority of cities and towns of Lithuania, Latvia, Estonia, Ukraine, Belarus', Tadzhikistan, Armenia, and Georgia is based on groundwater. Minsk, Tbilisi, Erevan, Alma-Ata, L'vov, Samarkand, Vilnius, Lugansk, Beshkek, Riga, Kiev, Dushanbe, Ashkhabat, Tashkent, Baku, Khar'kov, and many other cities satisfy their drinking-water requirements mainly or solely by groundwater (Borevsky et al., 1987; Zektser et al., 1987; Zektser et al., 1993).

Most significant groundwater potential reserves are in large artesian basins of the platform types and large folded areas. Large groundwater development areas are characteristic for intermontane depressions and foredeeps, characterized by favorable conditions for groundwater accumulation. Moduli of potential reserves here amount to 10 l per sec per 1 km^2 and more, and separate well-field yields exceed some cubic meters per second (the Ararat, Chuisk, Issyk-Kul' basins, Fergana valley, intermontane depression of Tyan'-Shan', etc.). Urban and rural populations' water demand is completely satisfied by the groundwater in these areas, which is mostly used and can be used in future for irrigation.

Favorable natural conditions for groundwater accumulation are observed in such platform artesian basins as Moscow, Dnepr-Donetsk, Baltic, West-Siberian, etc., where potential reserves moduli are from 1-2 to 3-5 l/sec per 1 km^2 and group well-field yields are measured by hundred liters per second, and in the river valleys to 1 m^3/sec. Most towns and settlements satisfy their demand for domestic and potable water with groundwater.

Most unfavorable hydrogeological conditions are characteristic for the Baltic, Ukrainian and Donetsk crystal shields, some areas of the North and South Urals, Siberia, Far North, Northeast, Far East, Central Kazakhstan and some others. Here, moduli of potential reserves usually do not exceed 0.1–0.2 l/sec, only in some well-watered sites they can reach 1–2 l/sec. Groundwater is mainly used here for water supply of rural and small-town populations.

What are the possibilities and prospects of increasing the efficiency of fresh groundwater use in Russia?

It is known that a general provision of a region with groundwater is characterized by potential groundwater safe yield, being a maximum possible groundwater withdrawal by well-field structures located over the entire area of aquifers occurrence, for a given estimated term of development and with a given drawdown value. A potential safe yield was determined virtually without allowance for environmental constructions, human impact, and technological and economic aspects of development. Potential groundwater yield should be considered the upper limit of possible development, which requires a further correction.

A real possibility for groundwater withdrawal by a group of water-supply wells depends on the value of explored safe yield assessed by federal or territorial commissions on mineral resources.

As follows from Table 2.5, the Russian Federation has an appreciable potential of groundwater safe yield that amounts to about 230 km^3 per year. Approximately half of this value (113 km^3 per year) can be included in the water-resources balance in addition to surface-

TABLE 2.5

Groundwater Safe Yield of Russia, km^3/year(Yazvin, Zektser, 1996)

Economic Region	Potential Yield		Assessed Yield		Groundwater Withdrawal	
	total	added to surface runoff	total	prepared for development	total	assessed resources
Northern	28.7	13.6	0.63	0.38	0.36	0.06
Northwestern	9.0	2.0	0.32	0.28	0.39	0.08
Central	19.2	10.5	5.81	4.44	3.41	1.74
Volga-Vyatka	8.0	4.6	1.46	0.71	0.50	0.17
Central Chernozem	9.5	4.5	2.38	1.85	1.43	0.79
Volga	11.0	5.9	2.30	1.12	1.37	0.28
North Caucasian	8.7	2.6	4.18	3.08	1.95	0.92
Ural	15.6	6.4	2.57	1.84	2.58	0.83
West Siberian	63.5	39.9	2.99	1.90	1.93	0.49
East Siberian	14.7	6.7	1.95	1.30	1.22	0.29
Far Eastern	40.3	16.2	1.93	1.12	0.74	0.29
Kaliningrad Region'	0.25	0.1	0.18	0.15	0.10	0.05
Total for Russia	228.5	113.0	26.70	18.17	15.98	5.99

water resources. Groundwater safe yield is formed as a result of a decrease in groundwater storage for 50 years and comprises groundwater that is discharged under natural conditions by evaporation from the water table or outflowed directly to seas and oceans.

As was mentioned above, providing all of the Russian territory with groundwater is characterized by the value of the so-called potential exploited reserves, being a maximum possible groundwater withdrawal, when well fields are distributed over the whole area of the aquifer occurrence. This value is about 230 km^3 per year in Russia, which exceeds its current use by 15 times and indicates great prospects of more-intensive groundwater use.

Explored storage is very slowly developed in Russia. Thus, at present only about 1 500 of the explored development areas are exploited, that is about a half of their total number, and only 1/3 of the groundwater storage, is approved and prepared for exploitation.

However, it should be considered that a given value of potential exploited reserves was determined without taking into account possible quantitative and qualitative changes of the resources caused by human activities, and without accounting for nature-protecting limitations, the necessity of which appears due to the impact of groundwater components. Consideration of these factors will result in some decrease of the groundwater potential-reserves total value.

The values presented show that the total groundwater abstraction, including mine dewatering and vertical drainage, accounts for about 7% of potential safe yield. The explored safe yield (26.7 km^3 per year), of which 18.2 km^3 per year are the reserves prepared for development, constitutes a substantial value that exceeds the existing groundwater withdrawal. If we compare the value of groundwater withdrawal from explored groundwater reservoirs and the value of reserves prepared for development, it becomes evident that the development of explored groundwater reserves is characterized by a small value (about 34% for the whole of Russia, with variations from 15.8% in the Northern Economic region to 46.3% in the central Chernozemny Economic Region). This small percentage of development can be partially explained by the fact that, in many cases, the explored groundwater safe yield is intended for meeting prospective water demands and by the incorrect strategy of water-management bodies.

It should be emphasized that groundwater pumping in areas with assessed reserves amounts to less than a half of the total withdrawal. This is because groundwater development by single wells and small groups of wells does not require groundwater-reserves appraisal. Moreover, some well fields withdraw groundwater reserves that have not yet been approved.

When estimating the prospects of groundwater development, it should be taken into consideration that safe-yield values are distributed in a highly irregular manner. For instance, the Northern Economic Region, Murmansk oblast region, and the Karelian Republic (located in the Baltic Fracture Water Basin) have scarce groundwater resources. Despite favorable recharge conditions (the recharge rate is estimated as 16 km^3 per year), the potential safe yield is equal to a mere 0.16 km^3 per year, which is due to poor seepage characteristics of the water-enclosing crystalline rocks. Vologda region is characterized by small seepage values and safe yield; groundwater of the western and southwestern parts of Arkhangel'sk region has an elevated dissolved-solids content. Most favorable conditions are typical for the central part of Arkhangel'sk region, where high-yield carboniferous fractured limestones and dolomites occur in the Onega-Severnaya Dvina interfluve area and in the area of the Komi Republic.

Different and generally insufficiently favorable conditions of groundwater development for drinking-water supply are also characteristic for the Northwestern Economic Region, especially for Novgorodsk region and most of the Leningrad region. However, this region has a vast area with highly favorable hydrogeological settings — the Silurian Plateau in the Leningrad region, where karstified and fractured limestones are water-bearing rocks.

The Central Economic Region has a substantial groundwater safe yield. Large groundwater reservoirs are formed in the Carboniferous and Devonian carbonate deposits of this region. Groundwater is used by many towns and settlements in the Bryansk, Vladimir, Tver', Kaluga, Moscow, Oryol, Ryazan', Smolensk, and Tula regions. The northern and eastern parts of the Central Economic Region (Yaroslavl region and some areas near Tver', Ivanovo, and Kostroma) have less favorable conditions. The potential groundwater safe yield in each of the Yaroslavl' and Kostroma regions is equal to a mere 0.35 km^3 per year; for this reason, centralized groundwater development there is extremely difficult.

A relatively small potential safe yield of fresh groundwater is generated in the Volga-Vyatka Economic Region; it is equal to 8 km^3 per year. Most of the region is located within the Eastern Russian Artesian Basin, where the main aquifers are often characterized by insignificant production rate or have water with an elevated dissolved-solids content. At the same time, individual areas of the region are highly promising for prospecting of even large groundwater reservoirs; these areas are the Tesha and Moksha interfluve in the Nizhny Novgorod region, the Mordovian Artesian Basin, Kirov Groundwater Reservoir in the Bystrisa River valley, and some other areas.

The Central Chernozem Economic Region is characterized by favorable conditions of the groundwater development for public water supply. Almost the entire system of public (municipal and rural) water supply is based on groundwater.

The North Caucasus Economic Region has diverse hydrogeological settings. In this region, the areas located in the Azov-Kuban' and East Ciscauscus artesian basins and in the piedmont valleys of Greater Caucasus (Krasnodar Territory, eastern Stavropol' Territory, Western Dagestan, Chechnya, Ingushetiya, Kabardino-Balkariya, and Northern Osetiya), are characterized by substantial groundwater resources, while a larger part of the Rostov region, west and central parts of the Stavropol' Territory, and Eastern Dagestan have unfavorable conditions for groundwater development. However, in these areas, prospective individual sites for centralized water supply using groundwater may be found.

Complicated and insufficiently favorable hydrogeological conditions are typical for the Volga Economic Region located within the East Russian and Caspian artesian basins. In this region, major productive aquifers are formed in the watershed zones of rivers and reservoirs, which makes it possible to construct high-rate well fields; however, these aquifers are vulnerable to contamination. Astrakhan' region and Kalmykiya are in the most difficult conditions: Centralized water supply using fresh groundwater is virtually impossible there. These areas are noted for the smallest potential safe-yield values (0.48 and 0.32 km^3 per year, respectively) and an insignificant groundwater percentage in the public water supply (3% in the Astrakhan region only).

Very complex and insufficiently favorable conditions for groundwater development for public water supply are also observed in the Ural Economic Region, although groundwater contribution is relatively large there. Groundwater is most intensely used in the region west of the Urals (in Orenburg region and Bashkortostan) where Quaternary aquifers are developed in the Ural, Samara, Belaya, and Ufa river valleys. Under these conditions, surface water is the main component of the safe groundwater yield, and because of this, the aquifers under development are vulnerable to contamination. An example of this situation is the Southern Well Field in the city of Ufa, where groundwater was repeatedly contaminated with phenol. In the Ural Hydrogeological Zone (the Sverdlovsk and Chelyabinsk regions), most important is the groundwater of small hydrogeological structures in intermontane areas composed of fractured and karstified limestones, particularly if surface runoff exists within the area. The Kurgan region, where mineralized groundwater widely occurs, is particularly lacking fresh groundwater.

The West Siberian Economic Region is located within a major artesian basin of the same name, in which substantial groundwater resources are generated. However, they are disturbed irregularly. Omsk region and the southern part of Tyumen' region have the least favorable conditions because they have mainly mineralized groundwater.

The East Siberian Economic Region is located for the most part in the permafrost zone; the thickness of the perennially frozen rocks is more than 600 m in the north of the region. In this connection, most-favorable conditions for groundwater development in this region are observed in river and lake valleys and in talik (thawed) zones. On the whole, groundwater is the principal source of public water supply in this region (in the Krasnoyarsk Territory, Chita region, and Tuva and Buryatiya republics); only in the Irkutsk region does surface water dominate.

The Far Eastern Economic Region is characterized by different availability of groundwater. Kamchatka region, Sakhalin region, Khabarovsk Territory, and Amur region are relatively rich in groundwater. The Amur River Valley, composed of alluvial deposits, and intermontane artesian basins in the Amur-Zeya and Sikhote Alin hydrogeological regions are most favorable for groundwater development.

Unfavorable conditions for groundwater development are typical of the Sakha Republic, Magadan region, and the Chukotka Autonomous District located in the permafrost zone.

Kaliningrad region is relatively poor in groundwater: its potential safe yield is equal to a value as low as 0.21 km^3 per year.

By characterizing a current situation with groundwater withdrawal, it should be noted that two groups of groundwater well-field systems for public water supply can be singled out in Russia: directly within urban territories (dispersed in many cases) and centralized well fields outside of them (Kochetkov, Yazvin, 1992; Zektser et al., 1993).

The first group incorporates mainly disordered systems of separate and small concentrated well fields consisting of some operating wells, elementally developed in urban territories concurrently with towns' development. These systems can be connected up to partially solve the problems of industrial or separate populated areas' water supply. In most cases, these well fields were drilled without special hydrogeological proving. The main problems of well field exploitation are sometimes connected with the actual impossibility of creating a sanitary protection zone around, resulting in groundwater contamination in this group.

The second group, depending on hydrogeological condition peculiarities and water-consumption volume, incorporates both large centralized well fields pumping out hundreds of thousands of cubic meters of water per day, and satisfying water demand of large towns or some small ones, and systems of relatively small well fields pumping out 10 000 cubic meters of water per day. These well fields are usually essentially removed from water consumers to a distance of 150–200 km and more. These well fields were designed and drilled on the basis of special prospecting and exploration work, calculated during this process groundwater safe yield has been approved by the State Commission for Mineral Resources Reserves. These well fields often have significant technical and economical advantages and are characterized by a higher groundwater quality, and mostly by protection from contamination. The future undoubtedly belongs to well fields of this group.

Analyzing the present state and prospects of fresh groundwater development over the area of Russia, it is necessary to discuss, roughly, a prediction of fresh groundwater development in a remote perspective, which is an important problem of public water supply. This problem is very complex and is virtually in the initial stage of its investigation. The complexity of its solution stems, above all, from the insufficient elaboration of methodological principles of extra-long-term prediction in this area, from a lack of data on trends in the development of individual industries and agriculture; new major industrial complexes, cities, towns, and other sites of water supply; and lack of scientifically substantiated standards of water use and requirements for the quality of ecologically pure and biologically valuable water for public water supply. Below, we will discuss only the main trends in changes in the safe groundwater yield of Russia and, consequently, public water supply, using groundwater as a remote perspective.

As was noted (Zektser et al., 1979), groundwater safe yield will be affected toward the middle of the next century by a multitude of factors that are primarily related to human impact. Let us discuss briefly the principal factors that bring both a decrease and an increase of the safe yield. A high-rate water withdrawal, hydraulic construction, development of mineral deposits, groundwater contamination, artificial recharge, and environmental-protection activities are major factors.

As was mentioned, unrenewable reserves that will significantly decrease during the next 50 years equal about half of estimated potential groundwater yield of Russia. Thus, if we assume that for the next 50 years, the total groundwater withdrawal will double and will equal about 35–40 km^3 per year, we may suppose that a total groundwater safe yield of Russia will decrease by about 15–20 km^3 per year as a result of unrenewable reserves.

Approximately half the large well fields are located in river valleys where groundwater reserves are mainly generated by seepage from rivers. Long-term regulation of surface runoff by reservoirs results in changes of the river-runoff regime, which in turn changes the regime of aquifer recharge in river valleys. The recharge of well fields that are located in the river banks is reduced when flood duration and intensity decrease.

At the same time, constructing of reservoirs and some other water-management measures (e.g., construction of canals and intensification of irrigation) are a positive factor (if groundwater is not contaminated) causing an increase in the groundwater safe yield that results from augmentation of the saturated thickness of aquifers and their recharge by seepage from canals, backwater, and infiltration from reservoirs.

As noted above, a significant amount of groundwater pumped out during the development of mineral resources and protection of flooded areas during irrigation and hydraulic construction is not actually used and is aimlessly discharged into surface streams.

Groundwater safe-yield decrease is also connected with contamination (Krainov and Shvets, 1987). However, available data do not permit us to reliably quantify this decrease. It is evident that the number of sites with contaminated groundwater is increasing. At the present time, abstraction of contaminated groundwater in Russia amounts to about 1 km^3 per year or about 6% of the total withdrawal. This value will probably increase somewhat in the future.

The necessity to observe the admissible rate of groundwater abstraction on other environment components will also provide for diminishing of the safe yield.

Artificial recharge, which may be used for regulating the decrease in storage, is an important factor in increasing the productivity of aquifers. Artificial recharge must be carried out first in the areas with operating well fields if conditions are adequate. Artificial creation of fresh groundwater lenses in desert regions is a similar action.

It is very important to improve methods and technological means of groundwater pumping. As is known, groundwater safe yield is now assessed with allowance for up-to-date means of development. There is no doubt that new types of pumping equipment will be elaborated and new constructions of high-production wells will be developed that will make low-cost groundwater withdrawal possible from a depth of about 400–500 m. This will permit development of deep aquifers whose resources are not taken into account at the present time.

A substantial reserve for water supply is brackish and saline groundwater, which may be desalinized or used in combination with fresh water. In Russia's southern regions, resources of this water are frequently comparable with those of fresh groundwater. If less-expensive technologies of desalinization of brackish and saline groundwater are elaborated and methods of its withdrawal are improved, a contribution of this water to a total water consumption will increase appreciably.

In recent years, many climatologists rather confidently predicted climate warming due to an increase of CO_2 concentration in the atmosphere that resulted from intense economic activities. The predicted anthropogenic transformations of climate will undoubtedly influence groundwater. Preliminary investigations show that changes in groundwater resources will proceed differently in different regions of the country. In southern European Russia, a decrease in groundwater recharge and, consequently, in natural groundwater resources, is possible, while in northern areas, the contrary is the case (Kovalevsky, 1993). However, it should be emphasized once more that the methodology of predicting changes in groundwater resources under the impact of possible climate changes and a general methodology of long-term prediction of changes in natural processes are as yet in the initial stage of development. Elaboration of pertinent prediction methodologies is an important problem of hydrogeology.

Analysis shows that the effect of various anthropogenic factors on groundwater resources may be both positive and negative. At the same time, we can assume that, with scientific substantiation of managing groundwater development, regulating its use, and replenishment, as well as taking some measures for contamination control, the total value of potential groundwater safe yield will be maintained at a present level and will be sufficient for meeting requirements for water intended for public water supply in a remote perspective in Russia.

2.4.4 Ecological and Hydrogeological Problems of Groundwater Use for Water Supply of Moscow

As was already noted, fresh groundwater is widely used for the water supply of urban populations in Russia. Potable water supply of most towns with populations not less than 100 000 is almost fully based on the groundwater. One-third of large towns with populations exceeding 250 000 are supplied with groundwater for drinking purposes, and another third with surface and groundwater jointly. However, the water supply of cities in Russia and primarily Moscow and St. Petersburg, with multi-million populations, is almost totally based on surface water.

It should be noted that, in recent years, different governmental, water-engineering, and nature-protecting organizations in Russia, and primarily in Moscow and the Moscow region, as well as leading projecting and research institutions, actively discuss the problem of increasing water supply of the capital both now and in the future. Deterioration of surface water quality used for potable water supply makes the problem very urgent. Here, great attention is paid to assessing and proving possibilities for a wider use of fresh groundwater in the Moscow region for providing population with drinking water. Up to the present time Moscow is one of the few cities in Russia supplied only with surface water. Due to this there is a constant danger of a water-pumping disorder resulting from possible damages that could cause surface-water pollution. Thus, use of protected-from-pollution fresh groundwater will inevitably increase the reliability of domestic and potable water supply for the city.

Let us consider groundwater use for the city of Moscow in detail, especially as ecological aspects of intensifying the groundwater use in this area causes troubles to its inhabitants.

Groundwater pumping in Moscow and the Moscow region began almost 300 years ago. It was significantly intensified by the end of the last century, when fresh groundwater of confined aquifers in carboniferous deposits began to be pumped out. At present, there are more than 10 000 water-pumping wells considered to be within the limits of the Moscow region. Observations of the level regime and groundwater composition are made by 1100 wells of the observation net. In the mid-1980s a computer geoinformation model, being a prototype of the present Geoinformational System (GIS), was designed for the whole Moscow region, allowing us to assess regional changes in hydrogeological and hydrological situations under changing water withdrawal volume and regime. This model includes observation data for more than 10 000 wells. Its use allows us to predict possible changes in the river runoff under increasing or decreasing groundwater withdrawal in the near-river well fields.

Fresh groundwater used for potable-water supply occurs at a depth of 300–350 m in the Moscow region, and in some places in the south and north that are confined to marginal parts of the Moscow artesian basin, the thickness of the fresh groundwater zone does not exceed 80–100 m. Natural fresh groundwater resources in the carboniferous deposits, characterizing mean perennial value of their recharge, are about 100 m^3/sec under mean annual groundwater discharge module, which is about 2 l/sec from 1km^2. Accounted groundwater withdrawal is about 50 m^3/sec on the average.

The geological-hydrogeological cross section of the Moscow region territory is represented by two hydrogeological floors: the lower one, constituted mainly by loams of Carboniferous age, and the upper loose sandy-clayey deposits of Cretaceous and Quaternary age. These water-bearing strata are subdivided by regional water-confined Jurassic clays with a thickness from 8–10 m to 30–40 m, which are often washed out in the river valleys. Resulting from long intensive exploitation, groundwater levels in carboniferous aquifers decreased by tens of meters (up to 80–100 m in towns of Sergiev Pasad, Balashikha, Lyubertsy, Podolsk, Khimki, etc.). It should be noted that intensive groundwater withdrawal has not yet caused irreversible changes in landscapes and vegetation and decrease of local runoff (except a few small rivers).

According to Yu.O. Zeegofer's data, about 80% of groundwater withdrawal is carried out by city well fields, when ecological situation has worsened there in recent years. These well fields, particularly those located in Moscow and the suburbs, are functioning under a constant hazard of pollution.

Groundwater quality deterioration in these well fields, especially in the areas near Moscow, as well as the above given considerations about water provision for the capital, has caused a necessity for creating a special "General Scheme of a United System for Water Supplying Moscow and Moscow Region Using Groundwater." The possibility of using groundwater for domestic and potable water supply of the population in Moscow region is proved in this "Scheme."

The united system, incorporating four groundwater well field systems, will be created: the North, South, East and West ones, with a total groundwater withdrawal of 2.7 mln.m^3/d, correspondingly 860, 1200, 500, and 140 thous.m^3/d. The authors based the scheme on the following main principles:

1. Intensification of groundwater use in the Moscow region is the only — actually having no alternative — way for improving the quality and reliability of water supplying the capital of Russia and the nearest areas.
2. Total groundwater withdrawal in the Moscow region must not exceed the amount of its natural resources, i.e., it must not exceed the amount of its annual natural recharge (for a long period of time).
3. First, potable-water demand must be satisfied in towns of the Moscow region (about 5 mln.m^3/d). For this purpose, it is planned to use both already explored and exploited groundwater-development areas not already included in a united system (3.8 mln.m^3/d), and new ones in the four sites included in this system (1.1 mln.m^3/d).

4. Only a part of the groundwater resources remaining after satisfying the water demands of the Moscow region will be used for Moscow's water supply (1.6 mln.m^3/d).

While assessing perspectives of intensifying groundwater use, an interaction between existing and projected well fields was studied by mathematical models. It is suggested to use 2.7 mln.m^3/d of groundwater to be distributed between the four systems as follows: North — 0.8, South — 1.2, East — 0.56, and West — 0.14 mln.m^3/d.

Groundwater quality in the sites included in the united system corresponds on the whole to a potable-water standard stated in Russia, except an increased content of iron and manganese. Besides, a decreased fluorine content is observed in the Southern system. North and East groundwater development areas are reliably protected from possible pollution, but South and West ones must be disinfected, since they are poorly protected. Hydrodynamic calculations made by specialists indicate that groundwater quality will change insignificantly under exploitation and it will be possible to use it for potable-water supply.

When elaborating a "General Scheme of a United System for Water Supplying Moscow and Moscow Region, Using Underground Sources," considerable attention was paid to predicting ecological consequences of groundwater intensified use. In particular, the level-lowering effect on vegetation and landscapes was analyzed in the upper horizon, possible changes of the river runoff were predicted (particularly small rivers runoff), as well as a danger of polluting exploited aquifers due to pollutant migration and under changing hydrodynamic conditions of interaction between ground and surface water, and separate aquifers with each other. Here, the authors of the "Scheme" indicate, quite rightly, that the analysis of experience in exploiting groundwater well fields is primarily important when predicting possible impact of groundwater withdrawal on the environment. It has already been noted that perennial groundwater exploitation, causing a fall of groundwater levels in Carboniferous aquifers by many tens of meters, has not resulted in considerable and hazardous ecological consequences, except a decrease of runoff in a low water period in some places.

The effect of groundwater exploitation on small-river runoff manifests itself in two ways: Sometimes surface runoff decreases in some places (the Moscow River in the upper stream, the Istra in the middle one, Pakhra, Nerskaya, Nora, and some others) due to groundwater recharge by rivers and reduced groundwater discharge into river. In other cases, river runoff increases, if compared with a natural one (the Vorya, Torgosha, Pazha rivers) due to discharge of treated wastewater into rivers. The Klyazma River is an example, as its runoff has been decreased in comparison with natural flow in the stream above the town of Noginsk and has been increased in the streams below the towns of Noginsk and Electrostal. Mathematical modeling, taking into account seasonal regulation of groundwater recharge, indicates that a "loss" to small-river runoff was about 10% in a year of normal water content and 17–18% in a year of 95% water content. In some parts of such rivers, where runoff in the low water period is decreased by more than 25–30%, special measures will be needed, such as channel dikes, small-river recharge with groundwater in extreme situations, etc. Recent modeling, considering changes of seasonal groundwater recharge, made for the northern part of the Moscow region indicated, that decrease of runoff in a low water period (the Dubna River) in a year of normal water content is 12% and in a year of 95% water content it is 23.7%, for the Velya river they are 26% and 50% correspondingly, which will inevitably require special nature-protecting measures.

It should be noted that the problem of intensified groundwater use in the Moscow region caused unprecedented interest and anxiety in the population. Ecological aspects of groundwater use have never been so actively discussed by specialists and ordinary people in the former USSR. This is mainly because 1) it is the first time in Russia that so considerable a groundwater withdrawal is planned for the potable-water supply of Moscow, and 2) an intensified interest among the population in ecological problems and the use of nature, including the dangers of large-scale groundwater use.

Based on analyzing the available experience in exploitation, the author's preliminary conclusions about the insignificant effect of withdrawal on groundwater, first from the surface aquifer and, hence, on vegetation, have been, on the whole, proven to his satisfaction. However, this optimistic conclusion, which is very important for the ecology of every region, must be supported and proven by further tests and experimental studies. Creation of complex environmental monitoring is one of the most important projects for the effective use of groundwater. Work within the scope of this monitoring will make it possible to determine conservation measures and their role in minimizing any possible negative effects of considerable groundwater withdrawal on small-river runoff, vegetation, karst-suffusion processes initiation and intensification, and the quality of pumped groundwater. In addition, the results of such work will make it possible to elaborate on scientifically based recommendations for assessing the ecological consequences of groundwater withdrawal on the environment that can be used for solving similar problems in other regions.

3

Principles of Regional Assessment and Mapping of Natural Groundwater Resources

In recent decades the investigation of regional assessment of natural groundwater resources and flow have been widely developed. This is for two main reasons: First, the necessity and ever-increasing need for determining perspectives of groundwater use in different regions that must be considered in regional schemes and projects for complex use and to protect groundwater resources. Second is the development of techniques for regional assessment of groundwater flow that will make it possible to objectively and economically assess natural groundwater resources by analyzing and handling the available hydrological and hydrogeological materials without making special expensive and labor-consuming explorations.

Natural resources are defined as rechargeable groundwater flow, characterizing the amount of its recharge by infiltration of atmospheric precipitation, filtration from rivers, and leakage from adjacent aquifers. Natural groundwater resources occur and are continuously renewed in the process of a total hydrological cycle. They are the upper limit that causes the recharge of constantly operating well fields with an unlimited term of exploitation, except for coastal types of well fields. A mean perennial amount of groundwater recharge (minus evaporation from groundwater level) is equivalent to groundwater flow, therefore, natural groundwater resources can be given in quantitative characteristics of groundwater flow. These characteristics include modules and coefficients of groundwater flow and coefficients of river recharge with groundwater.

A module of groundwater flow is defined as groundwater flow discharge from a unit of catchment area, given in liters per second per 1 km^2 and thus characterizing natural productivity of the aquifer being assessed.

The coefficient of groundwater is the ratio of groundwater flow to atmospheric precipitation. It demonstrates (usually on a percentage basis) what part of atmospheric precipitation recharges the groundwater.

The coefficient of river recharge with groundwater is the ratio of drained groundwater flow to total river runoff, and it characterizes a portion of groundwater in the river runoff. This shows (usually on a percentage basis or parts per unit) what part of the river runoff is formed by the groundwater.

Given quantitative characteristics (modules, coefficient of groundwater flow and river recharge with groundwater) make it possible not only to characterize natural groundwater resources, but also to consider them as important water-balance characteristics, allowing us to compare different components of total water balance and total water resources for different regions.

To cite one example: On the average, 600 mm of atmospheric precipitation falls yearly in the area of Moscow. A mean annual module of total river runoff for a many-year period is about 6 l/sec per 1 km^2 (equivalent to a layer about 190 mm per year). The groundwater flow module, calculated by the method of generic stream hydrograph separation for a many-year period is about 2 l/sec per 1 km^2, which is equivalent to a layer of 63 mm/year.

Thus, it is clearly seen that a groundwater flow coefficient is about 10% in the area of Moscow, therefore, one-tenth of atmospheric precipitation contributes to groundwater recharge. The groundwater portion in total river runoff, or, in other words, the relationship between groundwater resources (groundwater flow) and total water resources (river runoff) is 30% on the average.

Under regional assessment, the aim in obtaining a mean amount of natural resources for a unit square or the total amount for a basin of ground- or artesian water, is assessing the area of the aquifer's occurrence or part of it.

At present, the main and most widely used methods for regional assessment of groundwater resources are the following:

- genetic stream hydrograph separation for a many-year period, hydrodynamic method for calculating groundwater discharge (modeling included)
- assessment of river recharge with groundwater by a change in their low-water runoff
- calculating the water balance of groundwater recharge or discharge areas etc. (Table 3.1)

Having no way of considering in detail methods for regional assessing groundwater flow and groundwater natural resources, and, as extensive literature is devoted to them (the main publications are given in references), the principal aspects of the two most commonly

TABLE 3.1

The Main Methods for Regional Assessing Natural Groundwater Resources

Methods	Advantages	Disadvantages
River hydrograph separation	Possibility of obtaining average long-term groundwater flow characteristics Possibility of evaluating groundwater flow variability	Need for long-term observations of a river runoff under disturbed conditions Applicable only to the upper hydrodynamic zone where groundwater discharges into rivers
Computation of changes in the river low-water runoff between two hydrometric stations	Possibility of obtaining both average long-term and annual and seasonal groundwater flow characteristics	Difference in the river flow between two section lines should exceed a total error in the river flow measurement.
Hydrodynamic method of computing a specific groundwater flow (analytical approach or modeling)	Possibility of evaluating groundwater discharge in individual aquifers	Need for good aquifer parameters, difficulty in averaging them. Impossibility to evaluate long-term groundwater flow variability
Method for determining a long-term water balance in groundwater recharge or discharge areas	Possibility of evaluating a discharge of deep aquifers not drained by rivers	Need for determining the main water balance components by independent methods Estimated groundwater flow value should exceed the error in determining main water balance components
Computation of infiltration values using groundwater level regime data	Possibility of evaluating groundwater discharge of individual aquifers.	Difficulties in areal extension of groundwater recharge values computed for a point (well) Need for numerous observation wells

used methods will be given, namely, the methods of stream hydrograph separation and the hydrodynamic method for calculating groundwater flow discharge.

The method for stream hydrograph separation according to generic types of recharge is based on a commonly held assumption that groundwater flow for a zone with intensive water exchange in the areas with a constant river system is formed mainly under a draining impact of the river system. Singling out the groundwater component in a total river runoff allows for assessing the amount of regional groundwater flow. Two parts are singled out under a genetic hydrograph separation: surface-slope runoff, intersoil runoff, and precipitation falling directly on the water surface in the river beds are related to the first part; river recharge with groundwater (often called a base runoff) is related to the second one.

At present, there are many scientifically proven methods and technical procedures for generic hydrograph separation. In this case, most authors proceed from the fact that stable low water level is formed only due to the groundwater flow (excluding rivers with prevailing lacustrine or swamp recharge). The absence of summer precipitation and winter meltings during the term, exceeding the time of high water wave passing through calculated hydrometric section, is a necessary condition for singling out a stable low water level.

The main difference in the available techniques concerns hydrograph separation during floods and high water. The approaches used here can be conditionally subdivided into three groups:

- not considering the effect of coastal control during a flood
- reducing the effect of coastal control to insignificant lowering of the river recharge with groundwater
- stopping groundwater discharge into the river as a result of coastal control

The experience in genetic hydrograph separation speaks for a necessity to consider concrete hydrogeological conditions of interaction between surface and groundwater, that is, a degree of their hydraulic connection.

Semicalculated methods for separation of the common river runoff hydrographs characterizing river-basin peculiarities are given in hydrogeological literature (Linsley et al., 1962; Wist, 1969; Chow, 1964; Freeze and Cherry, 1979).

To obtain reliable data on river recharge with groundwater, it is necessary to jointly consider the surface and groundwater regime of runoff within a catchement area, and prove the character and degree of their interaction. The processes of coastal control during floods cause a considerable decrease and sometimes stop river recharge with groundwater, which should be considered under hydrograph separation.

Russian specialists have developed a complex hydrologic–hydrogeologic method for hydrograph separation that was successfully used for regional assessment of groundwater discharge in the USSR territories and countries of Central and Eastern Europe (Map of Groundwater Discharge in the USSR, 1965; Map of Groundwater Discharge in Central and Eastern Europe, 1983; Kudelin, 1960; Zektser, 1977; Lebedeva, 1972; Dzhamalov, 1973; Vsevolozhsky, 1983; Groundwater Discharge …, 1982).

The main feature of this method is to consider the character and degree of interconnection between ground- and surface water in the river basin, which are determined as a result of careful study of the available geologic–hydrogeological data. In difficult cases, a reconnaissance or special investigation of the river valley is carried out. A typical scheme of draining for different parts of the river basin are made based on literature and field data. Drained aquifers and their lithological composition, and also levels of ground- and river water for different seasons are given in these schemes. The character of hydraulic connection between a river and aquifers, depending on the relationship between levels of ground-

and river water, causes the regime of groundwater flow into the river and defines different schemes for hydrograph separation.

The simplest way to assess groundwater flow drained by the river is to calculate low-water-runoff changes for a many-year period in the river site between two hydrometric sections.

A hydrodynamic method for assessing groundwater discharge is based on studying hydrogeological parameters of the main aquifers. Here, maps of the level surface and transmissivity, indicating common regularities of groundwater discharge formation, are compiled for every aquifer. Total groundwater discharge is determined by the main Darcy dependence for flow paths singled out. This traditional method, being very simple, gives the possibility of obtaining a reliable enough value of groundwater for each aquifer. Here, special attention should be paid to the reliability and accuracy of the initial hydrogeological parameters, compiled on their basic hydrodynamic maps. Basing the analysis on hydrogeodynamic conditions, a number of rated sites with similar hydrogeological parameters are singled out. Within them, flow discharge is calculated by flow paths, taking into account all the main parameters. Initial hydrodynamic parameters are not averaged for large territories, but are specified and given in detail for calculated sites and flow paths.

In complex hydrogeological conditions, with enough available data characterizing regional conditions of groundwater filtration, different methods of modeling are applicable for groundwater-flow assessment. Methods for assessing the interconnection between aquifers of an artesian basin should be considered the most suitable for this kind of calculations. Under a known areal distribution of head and transmissivity, this allows the gathering of horizontal and vertical components for groundwater flow at every point of calculations (Ogilvi and Semendyaeva, 1972; Dzhamalov, 1973; Zektser et al., 1984).

Groundwater flow within a water catchment described by the models of total river runoff formation is of great practical value (Kutchment, 1983; Khublaryan et al., 1986). There are approaches proposing a complex consideration and solution of differential equations for moisture transport, groundwater filtration, and water flow in the river bed (San Venan equation).

The other approach is based on using the equation for saturated–unsaturated filtration, which allows decreasing of equations considered. One method of modeling groundwater flow based on equations for depletion appears worthy (Greenfield, 1978).

The main methods for regional assessment of groundwater flow, their advantages and disadvantages are given in Table 3.1. Thus, for instance, a widely used method, primarily in the territories of sufficient humidity, for determining groundwater flow by generic stream hydrograph separation along with important advantages (the possibility of obtaining mean perennial data to characterize groundwater flow variability for a long-term period) is essentially restricted. It is most important to use data of an undisturbed river runoff regime, the assumption of coincidence between water catchment areas for surface and groundwater (which is impossible for areas of intensive karst and fissured rocks distribution), and to use data for long-term observations. Each of the methods mentioned has both advantages and disadvantages. That is why the right choice will depend on concrete geologic–hydrogeologic and hydrologic conditions of investigated regions, and also on the aims and scale (details) of the investigations made. The given methods are not competing; they supplement each other very well. That is why the most reliable result is obtained using a combination of different methods to assess regional groundwater flow.

It should be noted that the first works for regional assessment and mapping of natural groundwater resources and groundwater flow were made in the former USSR territory at the beginning of the 1960s under the initiative and guidance of Professor B.I. Kudelin. His work resulted in the compiling and editing in 1964 of the "Maps of Groundwater Flow of the USSR Area" at a scale of 1:5 000 000 and a monograph entitled "Groundwater Flow of the USSR Area" (1966), which is actually a detailed explanation note to this map. Later, in the early 1970s, the maps of groundwater flow in the USSR at a scale of 1:2 500 000 were compiled by a large group of hydrogeologists and hydrologists. At the same time, many years work began to assess and map groundwater flow in Central and Eastern Europe. This work was carried out in accordance with the UNESCO International Hydrological Program and resulted in the "International Map of Groundwater Flow in Central and Eastern Europe" at a scale of 1:1 500 000, and a monograph entitled "Groundwater Flow in Central and Eastern Europe" (1983) (Figure 3.1). Here, values of groundwater flow for large regions have been obtained, the main regularities of groundwater-flow formation, depending on physical–geographical and geologic–hydrogeological conditions have been revealed, and time and space peculiarities of changes for specific values and coefficients characterizing groundwater flow have been defined. Considering the positive experience of international cooperation in the field of regional assessment and mapping of groundwater resources in the period from 1987 to 1992, in accordance with UNESCO's Project for the International Hydrological Program, investigations have been made for regional assessment and mapping of groundwater flow of the whole globe. A large group of scientists from many countries (former USSR, USA, France, Australia, India, Brazil, Argentina, Thailand, and etc.) participated in this work. As a result of their joint effort, the "Map of the World Groundwater Flow and Hydrogeological Conditions," at a scale of 1:10 000 000, was compiled, then edited and published by an international group of experts in the USA in 1999. Among other works on the problem under consideration, studies made in different years for the regional assessment and mapping of groundwater flow and groundwater resources of the Russian Nechernozemie, Moscow, and Baltic artesian basins, Eastern Siberia, Cis-Caucasus and other regions of the former USSR territory, should be noted. The "Map of Groundwater Flow in California" at a scale of 1:2 000 000, published in 1991, jointly compiled by Russian and American specialists, should also be noted (Figure 3.2).

While not going into detail on the content and legends for the maps mentioned and other published maps of groundwater flow, the most important thing should be noted: Regional quantitative characteristics of the main aquifers (groundwater modules and coefficients of the river recharge with groundwater) characterizing their natural productivity and groundwater recharge in natural conditions are given in these maps. These maps contain quantitative information on groundwater and its resources, which makes them different from other hydrogeological maps. Besides natural conditions, factors (mainly geologic–hydrogeologic) causing groundwater resources formation are given in the maps of groundwater flow.

Maps of groundwater flow are widely used in practice (hydrologic–hydrogeologic and water-management works), allowing practical problems for complex use and protection of water resources to be solved on a quantitative base. Such problems incorporate determining fresh groundwater natural resources for characterizing water supply of separate areas; determining and predicting changes of groundwater component for the river runoff; assessing the amount of groundwater recharge when characterizing its safe yield; quantitative assessment of groundwater flow as an element of water balance for the territories, etc.

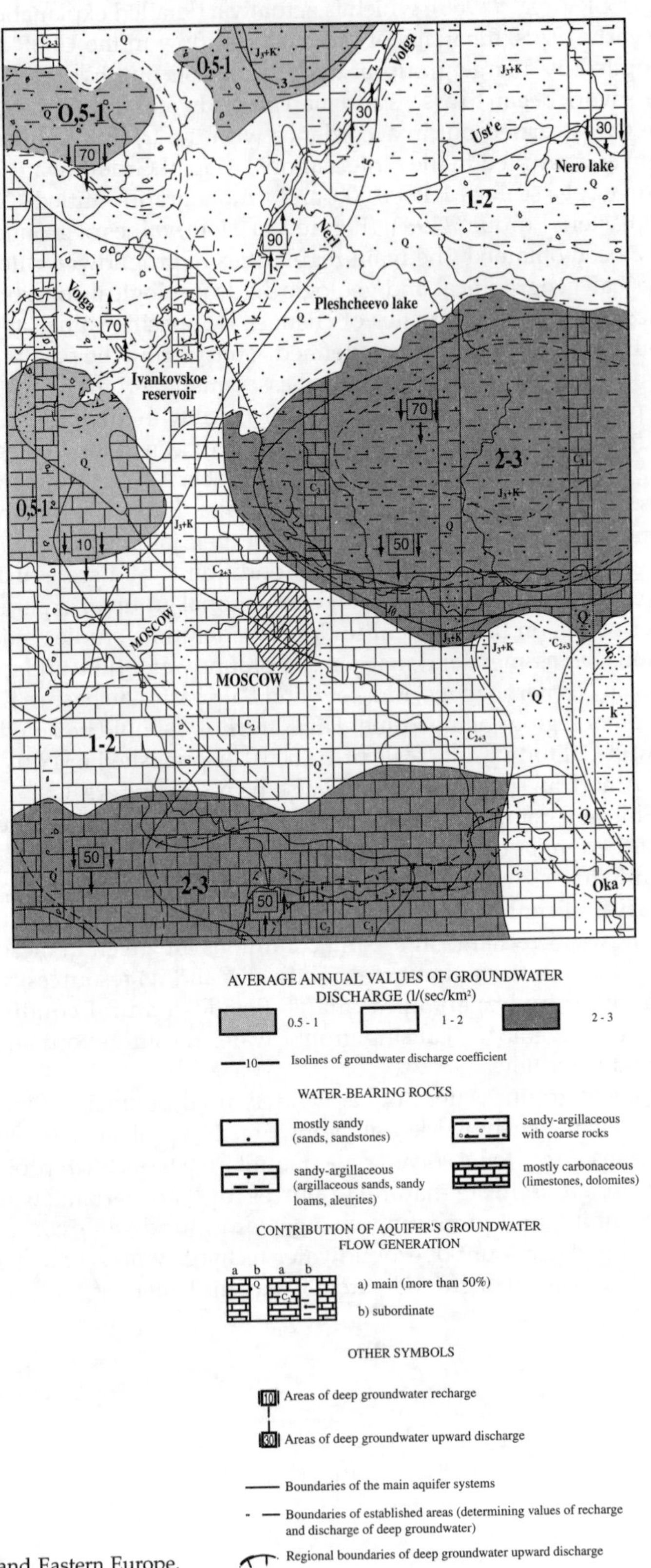

FIGURE 3.1
Groundwater flow in Central and Eastern Europe.

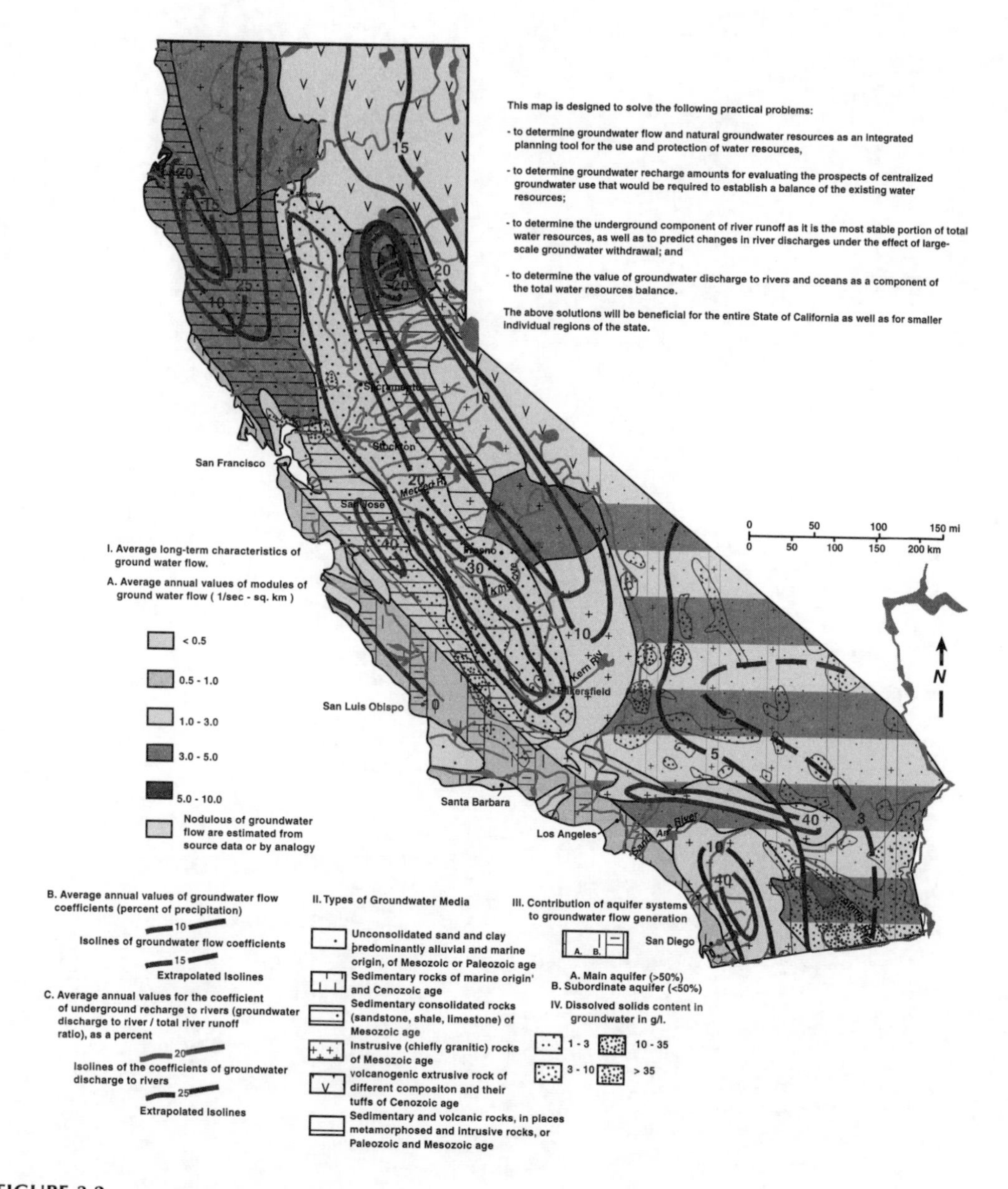
This map is designed to solve the following practical problems:
- to determine groundwater flow and natural groundwater resources as an integrated planning tool for the use and protection of water resources,
- to determine groundwater recharge amounts for evaluating the prospects of centralized groundwater use that would be required to establish a balance of the existing water resources;
- to determine the underground component of river runoff as it is the most stable portion of total water resources, as well as to predict changes in river discharges under the effect of large-scale groundwater withdrawal; and
- to determine the value of groundwater discharge to rivers and oceans as a component of the total water resources balance.
The above solutions will be beneficial for the entire State of California as well as for smaller individual regions of the state.
San Francisco
San Jose
Stockton
Fresno
Bakersfield
San Luis Obispo
Santa Barbara
Los Angeles
San Diego
Kern Rv
0 50 100 150 mi
0 50 100 150 200 km
N
I. Average long-term characteristics of ground water flow.
A. Average annual values of modules of ground water flow (1/sec - sq. km)
< 0.5
0.5 - 1.0
1.0 - 3.0
3.0 - 5.0
5.0 - 10.0
Nodulous of groundwater flow are estimated from source data or by analogy
B. Average annual values of groundwater flow coefficients (percent of precipitation)
10
Isolines of groundwater flow coefficients
15
Extrapolated Isolines
C. Average annual values for the coefficient of underground recharge to rivers (groundwater discharge to river / total river runoff ratio), as a percent
20
Isolines of the coefficients of groundwater discharge to rivers
25
Extrapolated Isolines
II. Types of Groundwater Media
Unconsolidated sand and clay predominantly alluvial and marine origin, of Mesozoic or Paleozoic age
Sedimentary rocks of marine origin' and Cenozoic age
Sedimentary consolidated rocks (sandstone, shale, limestone) of Mesozoic age
Instrusive (chiefly granitic) rocks of Mesozoic age
volcanogenic extrusive rock of different compositon and their tuffs of Cenozoic age
Sedimentary and volcanic rocks, in places metamorphosed and intrusive rocks, or Paleozoic and Mesozoic age
III. Contribution of aquifer systems to groundwater flow generation
A. B.
A. Main aquifer (>50%)
B. Subordinate aquifer (<50%)
IV. Dissolved solids content in groundwater in g/l.
1 - 3
3 - 10
10 - 35
> 35

FIGURE 3.2
Regional groundwater flow map of California. (Prepared by I.S. Zektser, L.G. Everett, S.J. Cullen, T.H. Robinson, and A. Macias.)

FIGURE 3.

4

Groundwater Contribution to the Water and Salt Balance of Lakes and Seas

4.1 Groundwater Discharge Into Seas: Present-Day Concept

Studies of groundwater discharge into the seas and oceans are part of a complex hydrologic–hydrogeologic problem of underground water exchange between land and the sea. Underground water exchange incorporates two opposite and non-equivalent processes: groundwater discharge into the sea and seawater intrusion into the coast.

This chapter is devoted to groundwater discharge into the seas, i.e., that part of groundwater that is being formed in the land and discharged into the seas, excluding the river network. Groundwater flux into the seas is formed in water-saturated coastal rocks, affected by the sea drainage. It occurs constantly everywhere, except in some areas of the Arctic and the Antarctic that are covered with permafrost of great thickness.

Studying seawater intrusion into coasts is an independent problem connected with analyzing the function of coastal well fields. Experience in theoretical and applied works in this field is mainly concentrated in the USA, Canada, Japan, the Netherlands, and France. Most international conferences and symposia are devoted to the problem of seawater intrusion into aquifers.

Seawater intrusion into the coasts under natural conditions is locally spread. However, this process is highly intensified under disturbed conditions. In many cases, a considerable groundwater withdrawal along the seacoast causes seawater inflowing (intrusion). Therefore, seawater intrusion into aquifers in some areas can be a serious threat to potable groundwater well fields.

It is noted in the work of renowned Spanish scientist E. Custodio (1982) that constant substitution of fresh groundwater with brackish and saline water under the impact of exploitation, particularly in submarine parts of deep aquifers, causes a change in groundwater balance that must be considered when calculating its reserves. An important aspect of investigations in this direction is studying the processes of physical-chemical interactions that occur under conditions of mixing ground and sea water of different composition.

Submarine discharge into seas and oceans is the least studied element of the present and prospective water and salt balance of the seas. First of all, it is because groundwater inflow is the only water balance component that cannot be measured, and data needed for a well-grounded calculation of a water balance underground component are often missing. Until recently, there was no technique for such calculations. At the same time, it is impossible to have a general picture of the world water balance and water balance of the oceans without data on the groundwater discharge.

The second reason for insufficient study of groundwater discharge as a component of water and salt balance of seas and oceans is subjective in character. For many years, hydrogeologists studying water balance thought that groundwater discharge was a small-by-value element of water balance if compared with its other components, and thus, it could be determined using the equation of mean perennial water balance. In other words, in their opinion, groundwater discharge can be determined as the difference between mean annual values of atmospheric precipitation and evaporation and river runoff. Thus, calculated groundwater discharge values totally depend on the accuracy of assessing mean values of precipitation, evaporation, and river runoff, and comprises all the errors of their estimation. In many cases, the authors in such calculations have groundwater discharge "mixed" with all the mistakes in determining the main components of water balance. This method of calculation yielded wrong conclusions of the water balance in the Caspian Sea where groundwater discharge values differed by more than 150 times according to the data of different authors made before 1970.

This approach is wrong in principle because groundwater discharge into the sea is usually small in comparison with other elements of the water balance (atmospheric precipitation, evaporation, river runoff). Therefore, it is important to determine it directly by hydrogeological methods.

Investigations of groundwater discharge into seas do not principally differ from that into big lakes, thus all the preceding and the following can also be applied to the water balance of lakes.

In recent years, practical needs concerning the problem of "interior seas" has given considerable impetus to development of research in the field of groundwater discharge into seas. The essence of the problem is that in many closed seas and big lakes there occurs considerable lowering of the water layers, caused by both natural factors and intensive economic activities in the catchment areas. The problem of studying the present and prospective water and salt balance of these reservoirs has appeared and hence, assessing the groundwater contribution to these balances formation. Here, groundwater impact on not only reservoir water and salt balance, but also on peculiarities of its hydrochemical, thermal, and hydrobiological regimes must be studied.

By now, considerable experience has been accumulated for quantitative assessment of groundwater discharge into closed and marginal seas and big lakes. These investigations are ongoing. Their main purpose is to study peculiarities and regularities in saltwater exchange processes between the land and reservoirs, and also to provide a prediction of changes in the groundwater component of the water balance under increasing human activities.

It should be specially noted that groundwater formed in the land and discharging in the coastal zone of seas and oceans, in many cases considerably affects hydrochemical, hydrogeological, and temperature regimes of seawater in the coastal zone and can affect the processes of sedimentation.

Nevertheless, groundwater discharge remains difficult to estimate and an insufficiently studied component of the water and salt balance of seas. Specialists must answer a number of complex questions:

1. What is the volume of this discharge?
2. Does it considerably affect the salt and water balance of the sea?
3. In what way will groundwater inflow change in the future under possible changes of climate and technological development in the coastal zone?
4. To what degree should the underground component be considered in studying the salt and heat balance in seas and oceans?

Hydrogeological problems related to groundwater discharge into seas have been principally considered. The main hydrogeological problem, closely connected with studying groundwater discharge into seas, or, in other words, groundwater exchange between land and sea, will be also given.

It has already been mentioned that many groundwater well fields along the seacoast and conditions of their functioning are mostly caused by ground- and seawater interaction. With sea- and groundwater interaction in coastal regions the aim is to determine the most optimal well field yield in the coastal region. As a result of intensive exploitation of such well fields, water exchange in the system "sea–ground water" changes. Hydrogeologists are faced with an important problem: determining the interface between fresh groundwater and saline seawater, and hence predicting water quality in the seacoast groundwater well fields.

Groundwater discharge into seas is an important indicator of groundwater resources. In coastal regions a deficit in potable groundwater of good quality can be considerably decreased or completely eliminated, using groundwater that now is "uselessly" discharging into the sea. Some countries have had positive experience in using large submarine sources, discharging into the sea in the coastal vicinity, and also in exploiting wells drilled in the shelf into fresh groundwater for water to supply coastal settlements.

Numerous cases of using water from submarine springs and separate wells drilled in seas for water supply are described in special literature. Separate examples will be given.

Submarine springs often occur on the underwater slopes of island systems with clearly manifested mountainous relief (the Hawaiian Islands, the Philippines, the Antilles, Big and Small Zond archipelagoes).

One of the world's largest submarine springs is near the coast of Jamaica, where a freshwater "river" was discovered with a yield of 43 m^3/sec. This source was discovered some 1600 m from the coast, with the water making its way to the sea surface from a depth of 256 m.

The Mediterranean Sea is rich with submarine groundwater outflows confined to fissures and karst channels in the rocks. In the Aegean Sea near the southeastern coast of Greece, a submarine freshwater spring with a large yield has been discovered. Near the coast of the Adriatic Sea there are about 700 submarine springs.

The Mediterranean Sea's springs are located at considerable depths, confined mainly to zones of local tectonic disturbances and karst deposits (near Cannes, at a depth of 165 m; San Remo at 190 m; in the Saint Martin Gulf at 700 m). In Dinaric, a seaside karst province at a distance of 420 km along the coast line, 32 outflows from separate and grouped submarine springs were revealed.

In the Mediterranean Sea, submarine springs are often so significant that they form freshwater flows. Thus, in Rome's river estuary, submarine sources were discovered on the sea bottom, forming a freshwater flow among saline water. A similar freshwater "river" also flows in the Gulf of Genoa.

Detailed studies of submarine springs, using different methods, have been made near the southern coast of France between the towns of Marseilles and Cassis (Figure 4.1). Port-Miou and Bestuan are the largest springs here. In 1964, the Bureau of Geologic and Mountainous Investigations and Water Society of Marseilles established a special scientific-research organization for studying these submarine springs and the possibility of using them for water supply, and to develop methods for studying springs of this type. It was determined that submarine sources are confined to karst limestones of Cretaceous age that form a monocline with a gradient toward the sea. Karst voids in these sediments manifest themselves at a depth of up to 100 m in the sea.

Divers investigating karst galleries of the Port-Miou sources reached a depth of 45 m below sea level and passed through the karst channels at a distance of more than 1 km inland. Observation posts were equipped with indicators of flow rate, manometers, and resistance meters; water and soil samples were taken; experiments were made with a coloring fluorescin, that enabled determination of filtration flow rate and direction; and geophysical experiments were made. It was determined that there is some water like mass layering in karst voids — saline sea water occurs at lower levels, fresh groundwater, having a lower specific weight, is above it. It was indicated that the rate of saline seawater flowing through the gallery is in inverse proportion to the head of discharging fresh groundwater. The value of the head in turn determines the yield of fresh submarine water flowing toward the sea by a dense seawater surface. The equilibrium between fresh and saline water is affected by the fresh groundwater discharge gradient and sea level fluctuations, and by the relation between fresh and saline water density and their temperature difference, thus facilitating the diffusion processes.

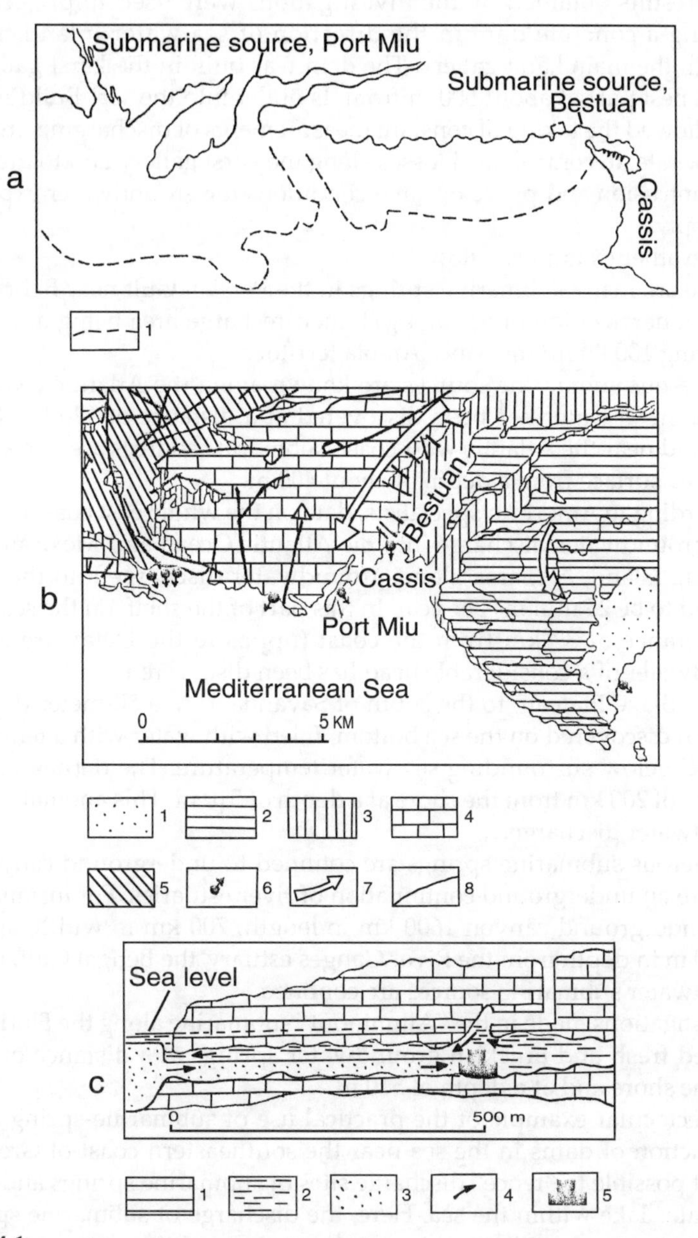

FIGURE 4.1
Groundwater discharge into Mediterranean Sea near Cassis (France). a: infrared thermal photograph of Port Miu and Bestuan submarine sources area; 1-boundary of fresh groundwater distribution in area. b: schematic hydrogeologic map of area; 1-alluvial sediments; 2-Senoman-Turonian limestone; 3-Upper Cretaceous rocks of marine genesis; 4-limstone of Urbomenian facies; 5-carbonate rocks of Lower Cretaceous and Jurassic periods; 6-submarine sources; 7-groundwater flow direction; 8-fault. c: scheme of karst gallery of Port Miu submarine source spring; 1-limestone; 2-groundwater; 3-marine water; 4-water flow direction; 5-dam. (Potieou, 1973)

The results obtained by the investigations were used in projecting and building a concrete dam in the direction of sea water inland intrusion through the main karst gallery. The dam was built in the karst gallery with the deepest part at about 500 m from its outlet into the sea. Building of this dam allowed the taking of constant measurements of discharging groundwater flow rate, to control head losses along the karst gallery and to prevent sea water intrusion and prove optimal conditions for groundwater exploitation for supply.

Some other examples follow.

There are many submarine springs in the Persian Gulf near Bahrain, with water mineralization of about 4 g/l, their recharge area being at a distance exceeding 100 km in the Saudi Arabia territory.

Numerous submarine sources are known along the Atlantic coasts. Near Florida, fresh water has been discovered at a distance of 120 km from the coast, and near the Atlantic west coast, submarine sources are known, rising to the sea surface from a depth of about 400 m.

According to American specialists, data on the whole east coast of the USA show groundwater discharge into the Atlantic Ocean and Mexican Gulf. In the Long Island, NY, area alone, groundwater discharge into the ocean is assessed to be 25 mln.m^3 per year. In this part of the shelf, on the sea bottom, at a distance of 37 km from the coast (opposite the Delaware estuary), groundwater of a considerable head has been discovered.

In the Blake Plateau, to the south of Savanna, GA, a 50-meter depression has been discovered on the sea bottom, filled with water with a temperature by 2.5°C below surrounding sea water temperature. The depression is at a distance of 200 km from the shore at a depth of 510 m. This anomaly is due to groundwater discharge.

Numerous submarine springs are confined to underground canyons that often are an underground continuation of river estuaries. For instance, there is an underground canyon 1600 km in length, 700 km in width, and more than 70 m in depth from the River Ganges estuary, the Bengal Gulf, to which groundwater submarine sources are confined.

Investigations made in the "Aluminout" submarine along the Florida coast revealed fresh and brackish groundwater springs at a distance of 120 km from the shore and at a depth of 510 m.

A spectacular example of the practical use of submarine-spring water is construction of dams in the sea near the southeastern coast of Greece that made it possible to "fence" discharge sites of submarine springs and create a freshwater lake within the sea. Here, the discharge of submarine springs is equal to about one million cubic meters per day and the water of this "lake" is used for irrigation of lands in the coastal area (Subsurface Springs from Freshwater Lakes, 1973).

At present, modern technical ways and means for withdrawing "water under water" are being developed in some countries. In Japan, a patent has been obtained for a means of fresh groundwater withdrawal from a submarine spring on the sea floor. The authors of the patent suggest separating fresh groundwater from the source and from the marine water immediately on the sea floor. To accomplish this, a special assembly with completely automated indicators constantly measuring water saline content is installed. If water salinity exceeds the permissible value then water supplying the consumer is automatically stopped and water is discharged into the sea until its salinity content is in conformity with the consumer's requirements, which have been previously determined.

Italian specialists suggested using a special bell for water withdrawal from submarine springs. The bell is installed on the sea bottom covering the source. It is equipped with a protective valve that controls water discharge and content, if necessary.

High perspectives in the field of submarine groundwater use by sea well fields appear due to a significant development of drilling appliances and wells test pumping in the continental shelf, slope, and bottom of seas and oceans. Some wells bored in the Australian shelf, near the USA Atlantic coast, in the Mexican Gulf continental slope, and in other places, have discovered fresh slightly mineralized submarine water having considerable head. Thus, fresh water was discovered during drilling in the Atlantic Ocean near the Florida coast at a distance of 43 km from the shore to the east of Jacksonville. Water with mineralization of 0.7 g/l was discovered when boring a well from a ship at a depth of 250 m below the sea level, where the water head reached 9 m above sea level.

However, it should be kept in mind that using the submarine water directly in the sea is quite difficult. This is due to the complexity of capturing (equipping) submarine-source outlets on the sea bottom, the necessity and economic expediency of capturing (piping) the flow, and the technical difficulties of boring wells in the sea. It should be indicated that conclusions on the possibility of using submarine water in practice can be made only after carrying out special works to assess these water reserves, including technical and economical proving of its expedience.

Investigations of groundwater discharge into the seas and large lakes are carried out in different countries with different aims and at different scales. Detailed studies of groundwater submarine discharge have been made for the Mediterranean Sea by French, Italian, Greek and Spanish specialists; for the Black Sea by Georgian, Ukraine and Russian specialists; for the Caspian Sea by specialists of some former USSR countries; for the U.S. Atlantic coast by American specialists. There are many examples of these detailed studies for separate areas.

For instance, according to data of Bodelle and Margat (1980), the submarine groundwater discharge from the area of France amounts to about 1 km^3 per year. The computations of the groundwater balance, made by modeling, for the Aquitaine basin (in the region of Bordeaux) show that the groundwater discharge to the ocean accounts for 10–15% of the total outflow components of the Eocene-Paleogene aquifer balance. According to the data of American researchers, groundwater discharge is observed all along the coasts of the Atlantic Ocean and the Gulf of Mexico. In only one area of Long Island groundwater is discharged to the ocean at an estimated rate of 25 mln. m^3 per

TABLE 4.1

Groundwater and Subsurface Dissolved Solids Discharge to the World's Oceans

Oceans	Groundwater Discharge *km³/yr*	Subsurface Dissolved Solids Discharge *millions T/yr*
Pacific	1,300.3	520.5
Atlantic and Mediterranean	815.3	470.3
Indian	219.4	295.5
Arctic	47.5	7.2
Total	2,382.5	1,293.5

year. The results of estimating the groundwater discharge to the Caspian Sea, Aral Sea, Baltic Sea (from the area of the former USSR) as well as to some major lakes of the former USSR (Baikal and Balkhash) are presented in the book by Zektser et al., (1984) (Table 4.1).

In recent years, investigations have been made by American specialists in a qualitative assessment of groundwater discharge into Lake Michigan.

It should also be mentioned that drilling in the bottom of seas and oceans has made possible the determination of considerable mineral reserves, primarily oil, gas, coal, iron ores, manganese, and phosphorites. A new problem to be solved by researchers is studying the groundwater contribution in the formation of mineral deposits on the sea bottom. These works are now only at the initial stage. Valuable conclusions on the origin of mineral deposits on the bottom of seas and oceans can be obtained when studying groundwater discharge into seas and the processes of physical-chemical interaction between submarine groundwater, rocks and seawater. The preliminary results of the work in this direction indicate that in the groundwater discharge areas on the sea bottom there occur abrupt changes of redox conditions that can cause significant phase transformations.

Thus, in the area of fresh groundwater discharge in the Black Sea near Gantiadi, the content of lead barium, nickel, chromium, vanadium, and zirconium in the bottom sediments is more than twice as much as background concentrations, which is due to the processes of interaction between discharging groundwater, bottom sediments, and seawater, in the author's opinion (Batoyan, Brusilovsky, 1976).

It has been noted above that groundwater discharge into seas and oceans occurs almost everywhere except the Arctic and Antarctic regions. The latest studies made in Antarctic regions indicate that there is a huge volume of fresh water under the ice. According to V.M. Kotlyakov, et.al's data (1977), there occurs bottom ice melting under the glacial cover, resulting in the yield of about 380 km³ of fresh water per year. The availability of groundwater is confirmed by geophysical and drilling works. Thus, fresh groundwater with a head of 60 m was discovered under the ice thickness by a well bored in the Berd observation station, Antarctica. This melted water discharges at great depth into the ocean and causes a desalting effect along the whole coastal region.

According to K.S. Losev's data, about 100 m³ a second of fresh water discharges into the ocean in Antarctica from underneath the Lambert continental glacier, which is confirmed by the availability of freshwater sediments in the underwater canyons.

According to R.K. Kligge (1995) groundwater discharge occurs not only in Antarctica, but also in Greenland and near some Arctic Isles.

Kligge assesses approximately that discharge in Arctic regions must be as much as 30 km^3 per year, and the total discharge of subglacial water into the oceans in the Arctic and Antarctic regions, taking into account V.M. Kotlyakov's data given above, is 410 km^3 per year.

According to the characteristic of its getting into the world's oceans and its contribution to the water balance, this subglacial water can be conditionally related to underground discharge, more precisely to an underground companion of the ocean water balance, though it is evident that origin and formation of the "subglacial" discharge into the World's oceans is principally different from groundwater formed in the land and discharged into seas and oceans.

Given the above considerations and examples we can formulate five main directions of research in the field of studying groundwater discharge into seas and underground water-exchange processes between land and sea.

The study of groundwater submarine-discharge sources is of great interest for clarifying peculiarities of marine bottom geological structures, revealing tectonic disturbances, and outcrops of more ancient rocks.

Qualitative assessment of salt exportation with groundwater into seas and oceans allows for characterizing groundwater denudational activities, i.e., for estimating the scale of coastal territories' surface decrease as a result of submarine groundwater discharge.

The main problems that face investigations of submarine groundwater, have been formulated. Even this short consideration explains the interest in these works. Scientific and practical importance of studying the process of underground water exchange between land and sea was repeatedly noted in special literature (Kudelin et al., 1971; Dzhamalov et al., 1977; Zektser et al., 1984). At the present stage of scientific development, five directions of research in this field and different aspects closely connected with each other can be formulated, taking into account concrete practical needs.

These directions are:

- the study of groundwater contribution to a total water balance of the earth and global water circulation
- the study of groundwater impact on the formation of water and salt balance in seas and oceans
- the study of marine and groundwater interaction in the coastal zone
- the study of fresh groundwater discharge areas in the coastal zone with the aim of using it for water supply
- the study of groundwater influence on mineral-deposit formation in the bottom of seas and oceans and the contribution of submarine-groundwater discharge to geological processes

Thus, it can be concluded that at present a new branch of science called marine hydrogeology is being formed. This is an independent part of common geology closely connected with marine geology, continental hydrogeology and oceanology, hydrochemistry and some other sciences. Submarine groundwater, or, more commonly, submarine hydrosphere, its properties, composition, structure, and processes of formation is the subject of marine hydrogeology research. At this stage of development, marine hydrogeology thoroughly investigates underground water exchange between land and sea, one of the elements of common water circulation in nature.

4.2 Methods of Study and Quantitative Assessment of Groundwater Discharge Into Seas

When studying submarine groundwater, conditions of its formation, flow, and discharge into seas, many different geologic, hydrogeologic, geophysical, aerocosmic, isotope methods, and methods specially developed for investigating processes of interaction between groundwater and marine water are used. The most common characteristics are given for the main investigation methods that make it possible to quantitatively estimate groundwater discharge into seas and oceans and to determine the location of submarine springs in the sea bottom are given below. These methods can be subdivided into two groups:

1. methods based on quantitative analysis of conditions for forming groundwater discharge into the sea within a catchment and primarily coastal areas of the land
2. methods of marine hydrogeological investigations based on the direct study of the freshwater area of the sea

The first group involves the analysis of geologic and hydrogeologic conditions of the sea coastal zone, which include a hydrogeodynamic method for calculating flow discharge (analytically and by modeling), a complex hydrologic–hydrogeologic method and a method of mean perennial water balance of groundwater recharge areas. These methods are primarily based on studying the processes of groundwater movement (flow), using a direct quantitative assessment of groundwater discharge into seas within a geological profile with available hydrogeological parameters.

The second group includes the methods for prospecting and investigating different anomalies in the sea water or bottom sediments that result from submarine groundwater discharge (anomalies in temperature and sea water composition of the bottom water layer, etc.). These methods permit areas of submarine groundwater discharge to be singled out and quantitatively characterized and in some cases make it possible to calculate the value of groundwater discharge, causing these anomalies.

Methods of remote sensing for detecting sources of submarine-water discharge also belong to this group, as well as methods for determining the rates of groundwater filtration into the sea through bottom sediments. Different modifications of special devices known as "seepage meters" and also various indicator methods, including isotope ones, should be related to the latter.

4.2.1 Methods for Studying Coastal Land

As was mentioned, submarine groundwater discharge occurs both in the form of concentrated springs along tectonic breaks and in the areas of fissured and karsted rocks, and by leakage through the poorly permeable roof of aquifers and marine bottom sediments. In this case, leakage processes often determine groundwater dynamics in the artesian basin on the whole or in its larger parts, and their function is essential and a requirement under submarine-groundwater discharge.

4.2.2 Hydrodynamic Method

The main method for quantitative assessment of submarine-groundwater discharge is hydrodynamic. Its essence is that the analysis of geologic-structural and hydrogeological conditions of artesian basins coastal zone, aquifers (aquifer complexes) are singled out, and the discharge is directed immediately into the sea, bypassing the river system. Using the designated aquifers, maps of the water table and water transmissivity are compiled, characterizing common regularities of submarine-groundwater discharge formation. The latter is determined along the sea coastline by the main Darcy's calculation function along singled-out flow strips (line). Calculation of flow discharge is made for every singled-out aquifer or complex, using available hydrogeological parameters determined by wells. There is no need here to use the data from all the wells; it is enough to use the methods of mathematical statistics for determining the number of wells needed for calculation of flow discharge with a required accuracy.

When investigations are more detailed and based on the analysis of hydrogeodynamic conditions of the coastal territory, separate areas with similar hydrogeological conditions are singled out for calculation. Within these sites, flow discharge is calculated by flow strips (lines), taking into account all the main initial hydrogeological parameters, characteristic of a relatively narrow coastal line.

In recent years, different modifications of the computer program Mod Flow have been widely used, which allows a hydrodynamic network of filtration flow to be modeled (distribution of piezometric heads and water transitivity values) and vertical and horizontal components of the groundwater flow discharge directed into the sea to be quantitatively assessed. The Mod Flow program's wide use for different purposes, including groundwater discharge in certain areas of seas and lakes, has given good results. Groundwater leakage into seas occurs in that part of the sea where piezometric levels of an aquifer are so much higher than the sea surface that there is a head gradient sufficient for the upward filtration of submarine water. Thus, determining the area of most intensive groundwater discharge by leakage and distribution of submarine-groundwater discharge is of great importance. For singling out this area, there is a method detailed in R.G. Dzhamalov's work (1977). Its essence is the following: Based on the maps of flow distribution in the aquifer, piezometric profiles are made by some flow lines. These profiles are then extended into the seawater area, and for this purpose are approximated by theoretical curves of different types. Piezometric profiles are prolonged into the seawater level up to the seawater line (edge) as much as possible, i.e., up to the border of the possible groundwater-discharge zone, by leakage. This method makes it possible to compile a defensible map of an aquifer piezometric surface within a sea coastal line. Such maps can be used for quantitative determination of groundwater leakage directly within the water area.

This technique was used to single out coastal zones of submarine discharge of confined water in the Caspian and Baltic seas.

Cuban hydrogeologist A. Valdes describes the experience of assessing groundwater discharge in the karst area of the Isle of Pines. Here a fresh groundwater lens floating in the saline seawater was discovered as a result of drilling, sounding, and water sampling. Studying the lens form and size and also water level in the wells in different seasons indicates that groundwater level decline in the period of infiltration recharge deficiency occurs due to groundwater discharge into the sea and total evaporation. As evaporation was insignificant, a change of lens volume for this period was caused by groundwater discharge into the sea. According to Valdes' calculations, the amount was 4500 m^3/d. The method he used relates to hydrodynamic methods of assessing groundwater discharge. Systematic observations of groundwater salinization, level, precipitation, and evaporation are required for studying groundwater discharge in a small island.

4.2.3 Complex Hydrologic–Hydrogeological Method

Mean perennial values of groundwater discharge modules of the main aquifers discharging into the sea are determined by different means within a coastal zone and the sea basin under study. The main procedure for this is a generic river runoff hydrograph separation by types of alimentation with a groundwater constituent being singled out. Then, using the analogy method, values of groundwater-discharge modules are extended to similar hydrogeological condition areas in the coastal zone discharging directly into the sea, and bypassing the river system. Multiplying the groundwater-discharge module by corresponding areas of the coastal zone, the groundwater inflow into the sea out of the intensive water exchange zone is calculated. Dimensions of the coastal zone areas, where discharge occurs immediately into the sea, bypassing the river system, are determined by hydroisohypses or hydroisopietic line maps of corresponding aquifers or, if they are lacking, approximately by maps of local relief. In the latter case, it is assumed that the groundwater table in the intensive water exchange zone corresponds in general to the land surface relief.

This complex hydrologic-hydrogeologic method has other restrictions. First, it is used only for those regions where the river system that drains the main aquifers is well developed. Second, perennial river runoff measurement, allowing the use of runoff hydrograph separation, must be carried out in this river system. Third, using this complex hydrologic-hydrogeologic method, a researcher does not determine in many cases the whole groundwater discharge into the sea, but only the discharge from the intensive water exchange zone, which contains mainly fresh groundwater. These restrictions determine a certain proximity of the results obtained. The complex hydrologic-hydrogeologic method is successively used for small-scale approximate estimations of groundwater discharge into seas. In particular, it was widely used for assessment of groundwater discharge from the continents into the world's oceans, the results being recently given in the map of the world groundwater discharge at a scale of 1:10 000 000.

4.2.4 Mean Perennial Water Balance Method

The method of mean perennial water balance can be used for assessing groundwater discharge into seas from deep artesian aquifers having a clearly delineated (distinct) groundwater recharge area. For the recharge area, an equation of the mean perennial water balance is written and the value of deep infiltration, i.e., the part of precipitation spent for artesian water recharge, is determined by a difference between precipitation, evaporation, and river runoff. This method is temptingly simple; however, its use is limited in some circumstances. First, it is usable for calculating water discharge out of aquifers that are reliably isolated by confining beds from the above- and below-lying aquifers. There needs to be a certainty that the discharge of the assessed aquifers "reaches" the sea and is not spent for leakage into other layers. Second, this method can be reliably used when an assessed value of deep infiltration exceeds the accuracy of determining other constituents of the water balance equation (precipitation, evaporation, river runoff). In other words, a researcher must be sure that using this method assesses precisely a deep groundwater discharge and not a discharge "mixed" with errors (mistakes, inaccuracies) on determining precipitation, evaporation, and river runoff.

The methods of quantitatively assessing groundwater discharge into seas make it possible to determine, with sufficient accuracy, a total submarine discharge out of all calculated aquifers and to get an idea of its distribution in the coastal land and sea. Here, two important circumstances should be emphasized. First, these methods do not compete but complement each other, and the most reliable result is obtained using both. Second, adoption of the methods considered is based on using already available hydrogeologic and hydrologic information and does not require carrying out special expensive drilling, test filtration, and other experimental works.

4.2.5 Method of Marine Hydrogeologic Investigations

As was mentioned above, groundwater discharge into the sea causes different anomalies in the seawater and bottom sediments. Studying the anomalies makes it possible to reveal and often to delineate the sources of groundwater discharge and to give its quantitative assessment. Therefore, marine hydrogeologic studies include a wide complex of visual, remote sensing, geophysical, geochemical, and other works aimed at studying different anomalies directly in the sea (marine water and bottom sediments) and caused by groundwater discharge. Thus, marine hydrogeologic studies (more exactly "hydrogeologic studies in the sea") are aimed at studying hydrogeological interaction between ground- and marine water and underground exchange between land and the sea.

The main methods for marine investigations aimed at revealing, studying, quantitative assessing, and mapping submarine groundwater are given in Figure 4.2. These methods can be subdivided into 1) remote and visual (cosmic survey, aerial survey, visual observations, 2) methods of directly inspecting submarine springs (investigation by bathyspheres, divers, indicator methods, and flow metering), 3) methods of studying the near-the-bottom water layer (determining anomalies in chemical, gas, and isotope composition, electric conductivity and temperature of the bottom water layer), and 4) methods of studying bottom sediments, including filtration properties and bottom sediments determining the use of flow meters of different construction, seismic-acoustic and thermal-electric profiling at a bottom/water boundary, studying of pore water chemical, isotope, and gas composition, etc. It is not possible to describe each method in detail (the latter is given in the special literature) therefore, only a short description and examples of applying certain of the most widely used methods for studying submarine groundwater discharge, which allows groundwater inflow into the sea to be quantitatively assessed, will be presented here.

The best perspective for studying submarine groundwater discharge is using remote sensing methods, primarily multispectral and infrared survey of the sea's surface from a plane or spacecraft. Multispectral survey is made by highly sensitive radiation in different spectrums. The survey is based on the differences in the spectral reflections of sunlight by different objects. This method makes it possible to get a maximum light contrast when interpreting elements of the surface in the investigated territory using corresponding light filters.

The possibility of using remote sensing methods for studying groundwater discharge into the sea is based on the fact that submarine groundwater springs cause changes in the main marine water parameters (color, transparency, temperature, water surface, structure), fixed on aerial cosmic photo-

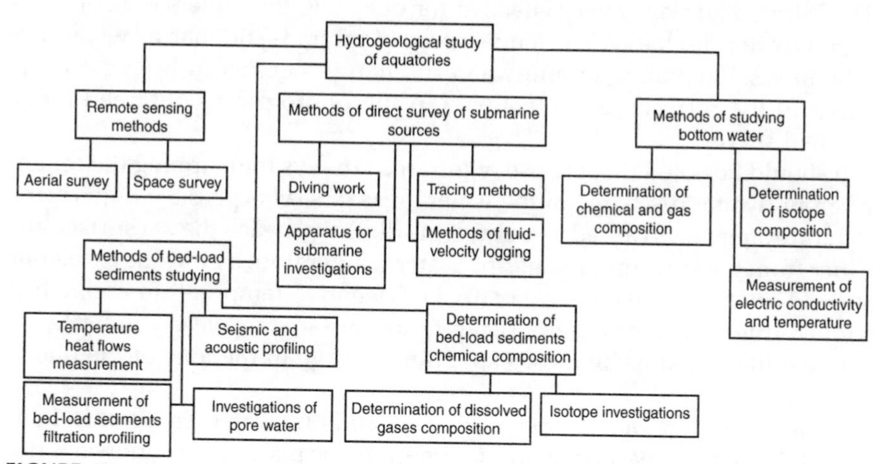

FIGURE 4.2
Methods of hydrogeological investigations of water area of seas and lakes.

graphs. Large-scale aerial photographs make it possible to study peculiarities of geologic and geomorphologic structure of the coastal sea and land in detail, and to single out faults and intensively fissured zones, to which large submarine springs are usually confined. A change of marine-water color or transparency caused by submarine springs is fixed on aerial photographs by more or less essential changes of a photo tone (shade) in a certain site that indicates a submarine spring. In some areas, groundwater springs are so abundant and their areas so wide (for instance, in the Al-Marj area in the Mediterranean Sea, where submarine springs are observed along the shore at a distance of 200 km), that anomalies caused by them are fixed even on cosmic photographs.

The first American satellite launched with the aim of studying resources of the earth carried a multispectral scanning device for registering electromagnetic radiation in four wavelength ranges: green and two closer to infrared. These ranges were not chosen at random, but were selected for different purposes. In particular, marine water is transparent in the "green" range, and therefore, bottom relief, turbidity, and different admixtures are seen in the cosmic photograph. Thus, large karst springs, where the water usually contains air particles or bubbles, can be detected. Similar springs can be observed on aerial photographs of the Jamaica coast, where upward karst water flows from turbid ovals on the sea surface. Multispectral surveys were used by the U.S. Geological Service to study submarine springs in the vicinity of Jamaica, Sicily, and Hawaii.

Infrared surveys are based on measuring the intensity of different natural surfaces' thermal radiation, including sea surface, in the infrared spectrum of electromagnetic waves. Using present-day infrared radiometers, it is possible to measure temperature changes on the sea surface with the accuracy of a 10th and even 100th parts of a degree by artificial satellites above the earth.

The infrared survey is very effective for detecting the centers of submarine groundwater discharge. The temperature of ground- and marine water usually differs. Submarine-groundwater discharge is observed by typical contrasts in the form of tails ("trains") in the photographs obtained by an infrared survey.

It should be noted that the use of infrared surveys for mapping the centers of groundwater discharge in the water areas of seas is possible only if temperature anomalies caused by submarine discharge reach the sea surface and differ from the surrounding marine water temperature by a value exceeding the radiometer's sensitivity. Positive or negative temperature anomalies, formed due to groundwater's rising to the surface, are clearly indicated in thermal photographs by a corresponding change in the tone of the photographs.

Infrared surveys were carried out when studying water resources in Long Island, NY, and, in particular, for determining the places of groundwater discharge in Long Island Sound. Here, the summer temperature of marine water is about 20°C and of groundwater, 10 to 15°C. Due to this, submarine discharge is distinctly observed along the shoreline by dark, hazy areas in the summer photographs.

Aside from detecting submarine springs, infrared surveys allow for a number of other important data to be obtained. Thus, submarine spring debit is approximately calculated, based on the area of the surface spot. The temperature of a spring can be determined by thermal photographs. Observations of both separate sites and large water areas are carried out by repeated surveys (particularly cosmic ones not requiring additional costs, highly periodic and covering wide territories).

Investigations of groundwater discharge by complex remote sensing methods were carried out in Big Quill Lake, a saline lake in the province of Saskatchewan, Canada. Nine centers of groundwater discharge with a total area of 4 km^2 have been discovered on the lake bottom by remote soundings, including aerial exploration by a reflected-wave method, cosmic surveys in four ranges from Landsat–1 satellite, and infrared aerial surveys. The combined use of remote sensing methods in this area made it possible to map with great accuracy the subaqueous groundwater discharge and to delineate the anomalies caused by groundwater discharge and flows in the lake.

Since ancient times, methods of determining subaqueous groundwater discharge have been known. Vigorous concentrated submarine springs cause different changes on the sea surface. Water domes have been noted where there is "boiling" of the sea surface, which causes a change of marine-water color because of gas bubbles. A change of color in the area of a submarine-spring discharge can be caused by turbidity and gas bubbles. Gas bubbles at the sea surface are caused by discharging the water of different origin. In karst areas, siphon springs can entrap air as they flow through the land. Discharge along the deep faults can contain water carrying methane, hydrogen sulfide, and carbon dioxide. Certain thermal springs carry gases of magmatic origin.

A change of marine-water color can be caused by different chemical reactions occurring under groundwater discharge on the sea bottom. A spring was discovered by a change of water color on fumarole fields of the Bannu-Vukhu volcano in the vicinity of the Indonesian coast. There water is red because of oxidation of iron brought by thermal submarine springs from considerable depth, while springs located at a shallow depth can be seen mainly in the coastal sea. Even effective springs can remain unnoticed if the depth of their discharge exceeds some tens of meters. Karst-spring discharge is subjected to essential annual fluctuations. In a dry season, their concentration decreases considerably and they become less visible.

When submarine springs are studied, immediate subaqueous observations using light diving equipment are very important. When hydrogeologic subaqueous surveys are conducted, such methods as morphometric description and application of different colorings and indicators are widely used. These methods include radioactive, chemical, and thermal sampling by different current- and flow meters, etc. of marine water along cross sections and profiles, filming, photographing, and measuring of a submarine-spring discharge and its hydrodynamic parameters. However, subaqueous hydrogeological investigations have some essential restrictions. The main drawback is the maximum permissible depth of diving in light diving equipment (40 m) when using compressed air (with an increase in diving depth, the time of a diver's stay under water decreases).

Investigations with light diving equipment were made along the Black Sea shelf in the area of the Gagry group of submarine springs with the purpose of studying the mechanism of groundwater discharge through karst springs on the sea bottom.

A detailed study of submarine springs using diving equipment was made near the southern coast of France in the area of Marseilles. Divers took samples of water and rock, and made morphological observations of submarine springs. They managed to move far along the karst channel and make subaqueous photographs.

Near the coast of Greece (in the Argolicos Gulf) at a distance of 400 m from the shore, divers managed to obtain water samples out of the submarine spring Aqualos at a depth of 72 m below the sea level. The spring, a narrow fracture, was described, and its discharge measured.

Description of hydrogeological investigations using diving equipment will not be complete without further experience in measuring submarine-spring debit and head by piezometers and flow meters of different construction.

By determining the head of groundwater, discharge through bottom sediments was made by special piezometers in Utah Lake, UT, where the locations of groundwater discharge into the lake water were discovered by infrared survey made from a plane at a height of 1100 m. These sites very closely coincide with anomalies of sources of sodium, magnesium, and potassium. Visual observations showed a triangular ice-free space with a base of 5 km along a coastal line in the discharge area. The triangular apex is at a distance of 3 km from the shore. Further, visual observations detected a great number of turbid swirlings 3 m in diameter, which are caused by groundwater discharge on the bottom. At the site of areal groundwater discharge, discovered on the lake bottom by infrared survey and aerovisual observations, a needle-filter, equipped with piezometer, is forced to a depth of 9 m. It should be noted that, in wells bored on the coast near the water line, groundwater level is above the level mark of the lake, and in piezometers in the lake water area, in the anomaly place the level reaches a height of 65 m above the lake level. This allows researchers to affirm that a discharge of confined groundwater is occurring out of the deep aquifers. When investigating Utah Lake, a combined set of methods for studying subaqueous groundwater discharge was used, including infrared surveys from a plane, aerovisual observation, hydrochemical investigation of the lacustrine water, and direct measurements of groundwater head on the lake bottom.

Flow meters can be used for assessing groundwater flow filtration through bottom sediments. Flow meter construction consists of enclosing a part of the bottom by a cylinder of a certain diameter, thus isolating a circle through which the subaqueoous discharge is measured. The open part of the cylinder is forced into the bottom sediments. The closed part of the cylinder, which has a small outlet, is connected to flow meters of different construction. By measuring the volume of water passing through a flow meter at a time unit from a certain circle of the bottom, a module of groundwater vertical discharge is obtained. By equipping this module with simple electronic devices, it is possible to make long-term regime observations. Using similar flow meters, a subaqueous groundwater discharge was studied in Lake Michigan, in glacial Lake Sally in Minnesota, in the Gulf of Mexico near Florida, and in Tanpo Lake in New Zealand.

Studying groundwater springs at a considerable depth is possible by using bathyscaphes, which make underwater photography and other observations possible over long time periods at any depth. Investigations made in the "Aluminout" bathyscaphe in the Florida shelf have revealed subaqueous fresh- and saline-water discharges and permitted their conditions to be determined. Investigations of the formation of hot brines in the Red Sea bottom and deep areas of Lake Baikal were carried out by Russian scientists in the "Pisis" bathyscaphe.

From the arsenal of methods for studying groundwater submarine discharge, different indicators are used. The method under discussion makes it possible to determine the location of spring-discharge conditions and, in some cases, to determine submarine-spring loss by dilution of the different indicators in sea water in the spring-recharge area or karst hollow. Indicator colored materials such as fluorescein, colored spores, different isotopes, and fresh spring water, were used to dilute the marine water.

To map and determine the loss of subaqueous groundwater springs, radioactive indicators are used. Serious research into submarine-groundwater discharge into the seas by isotope methods is being carried out in the U.S. at the present time (W. Moore, 1996; I. Cable, 1996; E. Kontar, W. Burnett, 1998). The technique of using flow meters (I. Cable, W. Burnett, et al., 1997) for determining groundwater filtration rates through marine bottom sediments (Cherkoner) was successively improved by American specialists.

Among the methods used on studying anomalies in marine-water composition and properties, the measurement of electric conductivity is widely used. This method is based on the relationship between dissolved-solids content (salinization), chemical composition, and the water's electrical resistance. Marine-water-specific resistance along profiles at different depths is measured by resistance meters used to detect groundwater springs in the sea. Temperature is registered simultaneously with measuring resistance based on the calculated water resistivity and salinity. Maps of marine-water salinity distribution are compiled, and the locations of submarine springs are determined by isoline configuration. The clearest results are obtained for effective concentrated karst springs.

In the detection of submarine springs, and, in some cases, for preliminary assessment of their water loss, anomalies of marine-water chemical composition are studied along profiles. These investigations are based on the essential difference between dissolved-solids content and chemical composition of marine water and groundwater.

Samples of marine water for chemical analysis are usually taken along some parallel profiles near the bottom, on the water table (surface), and in the intermediate depth. Similar methods were used by I.M. Buachidze and A.M. Meliva when studying karst submarine springs near the Caucasus coast of the Black Sea. As a result, 24 hydrochemical anomalies interrelated with submarine springs have been discovered. The maximum depth at which the anomalous chlor-ion content registered as different from the normal sea water by 5g/l was 400 m.

Similar surveys were carried out by different researchers in the Mediterranean Sea, along the Florida Coast, and in other areas. It is very convenient to use automatic analyzers on board a research ship that will allow the marine water to be continuously determined along a profile.

In recent years, the analysis of isotope composition of marine water was used for studying submarine-groundwater discharge. Changes in concentration of tritium, deuterium, radiocarbon, and oxygen in submarine springs enable opinions on their origin, recharge sources, water flow rate, and water-exchange terms.

Anomalies in bottom sediments are studied by different geophysical methods including temperature and thermal-flow changes, seismoacoustic profiling, thermal and electric profiling at a bottom/water border, and also by determining filtration properties of bottom sediments and analyzing chemical, isotope, and gas composition of their pore solutions.

A complex geophysical method (called a "marine hydrogeological complex" in special literature) was developed and used in Russia. The method consists of continuous seismoacoustic profiling carried out on board a research ship, and continuous measuring of temperature and salinity near the bottom layer by a special probe (Zektser et al., 1984, 1986). The essence of the method is in registering on the ship a geological structure of the bottom along a continuous profile by a seismoacoustic signal (the depth of seismoacoustic sounding reaches 1000 m if necessary). The temperature is registered along the same profile (with accuracy of 0.05%) at the "bottom/sea water" border. The results of profiling are continuously obtained in the form of graphs plotted by a recorder while the ship is moving.

Complex profiling is carried out on board research ships according to a chosen network. By applying this technique, a geological profile of bottom sediments and graphs of water temperature and salinity near the bottom along the profile have been obtained (Figure 4.3). Complex analysis of these materials makes it possible to reveal and delineate groundwater discharge centers on the sea bottom, and calculations based on the determined functions allow the value of submarine discharge to be quantitatively assessed.

The calculation of submarine-groundwater discharge, which causes temperature and salinity anomalies in the near-bottom layer, is based on studying salt and temperature balances and the anomalous sites where mixing of groundwater and marine water occurs.

The combination of geophysical methods was successively used when studying groundwater discharge into the Caspian Sea and Issyk-Kul and Balkhash lakes.

In conclusion, it should be stressed again that the joint use of several methods for assessing groundwater discharge into the sea increases the reliability of the results obtained.

4.3 Groundwater Contribution to Global Water and Salt Balance

Groundwater penetrates seas and oceans in three different ways:

1. as juvenile water, generated as a result of processes of degasification of the earth's mantle
2. with river runoff as base flow
3. as groundwater flow generated on land and discharging directly into the sea bypassing the river network

The inflow of juvenile water into the sea or ocean is a subsurface component of the water balance of seas and oceans, but, evidently, the term "groundwater discharge" is inappropriate here. This problem is very complicated, but is closely connected with the general problem of hydrosphere formation and water origin. Quantitative estimates of juvenile water inflow to seas are now difficult to make. It is assumed that the annual amount of juvenile water supplied by volcanoes, thermal springs, and deep faults is no more than one cubic kilometer, i.e., a value that is exceedingly small for the water

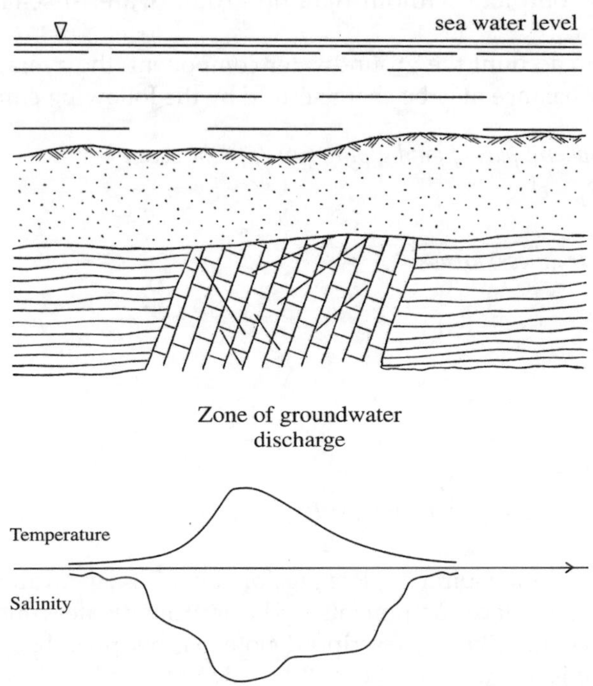

FIGURE 4.3
Salinity and temperature anomalies in seawater connected with groundwater discharge.

balance of oceans. At the same time, some amount of water is excluded annually from the hydrological cycle; it is above all the chemically and physically bound water of sedimentary rocks. In the opinion of most investigators, the amount of juvenile water coming from magma is approximately equal to the water amount excluded from the hydrological cycle as a result of sedimentation processes.

The second way is river drainage or groundwater discharge. The methods for calculating its base flow is well developed at present. This discharge gets into the ocean with river water, is part of total river runoff, and is considered in water balance calculations.

The third way is for groundwater to discharge directly from a continent into an ocean. This component is usually "lacking" under quantitative investigations of water balance. That is why groundwater discharge into the sea means the water formed in the continents and discharging into seas, bypassing the river systems. If we speak about total volume of groundwater in the seas and oceans, it should be recognized that about a third of the river runoff getting into seas is also formed due to groundwater drainage in the intensive water exchange zone.

As has already been noted, the lack of proved quantitative data on groundwater discharge into seas and oceans has restrained the studying of total water balance and water circulation up to the present time. World water balance was incomplete without data on groundwater discharge into the world's oceans.

Taking into account the groundwater component, the average long-term world water balance may be characterized by the following equations:

for the peripheral part of land giving runoff to the ocean

$$E_p = P_p - R_p - U_o \tag{4.3.1}$$

for closed (drainless) areas

$$E_c = P_c \tag{4.3.2}$$

for oceans

$$E_o = P_o + R_p + U_o \tag{4.3.3}$$

for the world

$$E = E_p + E_c + E_o = P_p + P_c + P_o = P \tag{4.3.4}$$

where E is evapotranspiration; P is precipitation; R is river runoff, including surface and subsurface components; and U is groundwater runoff from land to ocean bypassing rivers. Subscripts denote: p is peripheral part of land; c is closed area; o is ocean.

It should be noted that another term characterizing seawater intrusion into coasts is to be introduced into equations 4.3.1 and 4.3.3. However, this process under natural conditions of coasts is not considered as a component of the world water balance at the present stage due to its local character.

Similar water-balance equations were suggested by well-known hydrologist M.I. L'vovich and then were cited by many authors. However, the introduction of a new term Uo, groundwater discharge formed in a peripheral part of the land, not drained by the river system and, passing it by, directed immediately into the sea or ocean, is fundamentally new in the above equations.

Only a few works discussing groundwater discharge to oceans have been written. R.M. Garrels and F.T. Mackenzie (1971), proceeding from general concepts, evaluate total groundwater discharge to oceans to be 10% of total surface runoff. The first attempt to estimate this process on a global scale from the hydrogeological point of view was made by R.L. Nace (1967, 1970).

In Nace's paper published in 1967, the groundwater outflow to seas, arbitrarily set equal to about 4% of the surface runoff, is given to be 1600 km^3 per year. Three years later, R.L. Nace gave a value seven times smaller (up to 230 km^3 per year) and validated the calculations to some extent. He assumed the average thickness of the world's coastal saturated zone discharging groundwater to be equal to 4 m, water-bearing rock porosity 25%, groundwater flow rate 3 m/day, and the value of unit discharge 35 l/sec. Multiplying the value of 35 l/sec by the length of the assumed world coastal area equaling 200 000 km (the coasts of the Antarctic, the Arctic and Greenland, composed of perennially frozen rocks, are excluded) produces a value of direct groundwater discharge to seas amounting to 230 km^3 per year. R.L. Nace's choice of average hydrogeological constants is highly disputable. The thickness (4 m) of the zone where submarine groundwater discharge is generated is highly underestimated. Besides, if we consider the porosity value (25%) assumed by R.L. Nace and the actual groundwater flow rate (3 m/day), the seepage velocity would be 0.75 m/day, which is characteristic only of karstified and highly fractured or coarse detrital loose rocks (with a seepage velocity of 0.75 m/day, the coefficient of permeability should be 75 m/day and the hydraulic gradient, or flow slope, must be 0.01). An average value of rock porosity (25%) is highly overestimated, since this value is generally typical of sands. Thus, R.L. Nace's data cannot be considered to be sufficiently substantiated (Zektser and Dzhamalov, 1988).

An approximate estimate of groundwater discharge to oceans, obtained using the hydrodynamic method, is presented in the book *World Water Balance and Water Resources of the Earth* (1978). The authors of this book determined the specific rate of groundwater outflow to all oceans from the coastal zone of the upper hydrodynamic zone, assuming the following values of hydrogeological parameters:

- the average total thickness of all aquifer systems whose water discharges to seas is 200 m
- the effective porosity of rocks of this zone is 10%
- average hydraulic gradient is 0.005
- the coefficient of permeability is 10 m/day

These values have been selected from existing values for upper aquifers in different seaside areas. The specific groundwater flow rate per one kilometer of flow front, i.e., per one kilometer of the coastline of all oceans, amounts to about 10 000 m^3/day. The length of the coastline of all oceans except for the coasts of Antarctica, Greenland, the Arctic and some other permafrost regions, is equal to 600 000 km. Thus, total direct groundwater discharge from all continents to all oceans amounts to 2200 km^3 per year.

More validated and differentiated estimates of groundwater discharge to seas and oceans from individual continents to oceans have been obtained using the combined hydrological and hydrogeological method (Dzhamalov et al., 1977; Zektser et al., 1984; Zektser and Dzhamalov, 1988, 1989).

Application of the combined hydrological and hydrogeological method permit obtaining the value of groundwater outflow to seas and oceans from individual drainage areas of parts of land. The method was divided into separate drainage areas where groundwater discharged directly to the sea, bypassing the river network. The boundaries of groundwater drainage basins whose water discharges directly to the sea are commonly determined using potentiometric maps. However, such maps are absent for most of the coastal areas of seas and oceans. At the same time, it is known that drainage divides of surface water and the groundwater of the upper hydrodynamic zone coincide in the majority of cases. In this connection, drainage areas of groundwater discharge to seas were roughly determined from the world hypsometric map on a 1:2 500 000 scale. About 480 areas were selected for computations. The specific groundwater discharge values for these areas were taken from published maps of groundwater flow of individual continents, the former USSR area, and some European countries (Lvovich, 1974; Groundwater Flow Map, 1975, 1983; Margat, 1980) as well as from the World Groundwater Flow Map on a 1;10 000 000 scale being prepared at the Water Problems Institute, the Russian Academy of Sciences. These values were extended to hydrogeologically similar areas whose groundwater discharges directly to the sea. The total amount of groundwater outflow to the sea was obtained by multiplying the average specific groundwater discharge value by the size of the area.

When analyzing the values obtained, one should remember two assumptions. First, specific values of groundwater discharge to rivers were extended to hydrogeologically similar areas directly drained by the sea. Second, the groundwater discharge from the upper hydrodynamic zone alone is considered. Detailed studies of groundwater outflow to individual seas show that the main portion of groundwater discharge to seas is from the upper hydrodynamic zone. This is due to the more rapid water circulation in upper aquifers, favorable conditions for their recharge, and, as a rule, higher permeability of this zone. Therefore it should be considered that the contribution of deep aquifers in zones of slow and very slow water circulation to the total groundwater discharge to oceans is insignificant. Of course, the above two assumptions as well as the assumption of the coincidence of drainage areas of surface and groundwaters of the upper hydrodynamic zone, in a number of cases, reduce the accuracy of computations of groundwater discharge to seas for separate areas, but generally do not influence significantly the final estimates, which reflect correctly, in principle, the scale of the phenomenon under study (Zektser and Dzhamalov, 1988).

For each computation area, not only the total inflow of groundwater discharge to oceans, but also specific characteristics — the areal direct discharge to the sea from one square kilometer of the drainage area of land and linear discharge of groundwater per one kilometer of coastline were determined. This made it possible to compare computation areas, correlate groundwater discharge values with various natural factors, and reveal the general laws of generation of this process. The results of the computations for individual oceans and continents are presented in tables 4.1 to 4.5.

Thus, the total groundwater discharge to oceans, estimated using the combined hydrological and hydrogeological method, amounts to about 2400 km^3 per year, including 1485 km^3 per year from continents and 915 km^3 per year from major islands. The substantial figure of discharge from major islands (more than a third of the total submarine groundwater discharge) may be explained by a number of causes. Above all, the largest oceanic islands (New Guinea, Java, Sumatra, Sakhalin, Madagascar, West Indies, and some others) are situated in tropical and humid regions, where heavy precipitation creates

TABLE 4.2
Groundwater and Subsurface Dissolved Solids Discharge to the Atlantic Ocean

Continents and Islands	Groundwater Discharge			Subsurface Dissolved Solids Discharge		
	Areal Values L/s/1 km²	*Linear Values 100m³/d/km*	*Total Values km³/yr*	*Areal Values T/yr/km²*	*Linear Values 1000 T/yr/km*	*Total Values millions T/yr*
Africa	3.9	40.4	208.7	99.9	12.0	169.2
Europe	4.2	15.4	71.2	47.8	2.0	25.8
North America	4.6	31.9	219.4	74.6	6.0	112.2
South America	3.0	28.2	185.3	40.2	4.3	77.7
Major Islands	4.4	12.0	77.7	76.0	2.4	42.9
Total			762.3			427.8

TABLE 4.3
Groundwater and Subsurface Dissolved Solids Discharge to the Indian Ocean

Continents and Islands	Groundwater Discharge			Subsurface Dissolved Solids Discharge		
	Areal Values L/s/1 km²	*Linear Values 100m³/d/km*	*Total Values km³/yr*	*Areal Values T/yr/km²*	*Linear Values 1000 T/yr/km*	*Total Values millions T/yr*
Australia	0.2	3.7	16.4	28.4	5.5	66.7
Africa	0.6	5.1	22.1	38.7	4.1	49.0
Asia	1.7	10.7	65.3	97.2	7.1	119.2
Major Islands	5.1	27.7	115.6	84.7	5.3	60.6
Total			219.4			295.5

TABLE 4.4
Groundwater and Subsurface Dissolved Solids Discharge to the Pacific Ocean

Continents and Islands	Groundwater Discharge			Subsurface Dissolved Solids Discharge		
	Area Values L/s/1 km²	*Linear Values 100m³/d/km*	*Total Values km³/yr*	*Areal Values T/yr/km²*	*Linear Values 1000T/yr/km*	*Total Values millions T/yr*
Australia	1.1	4.6	7.1	24.9	1.2	5.0
Asia	4.8	27.2	254.3	98.2	6.5	165.2
North America	5.4	31.9	124.6	50.1	2.4	36.7
South America	11.5	58.7	199.6	64.1	3.8	35.5
Major Islands	13.0	51.0	614.7	159.8	7.3	278.1
Total			1300.3			520.2

TABLE 4.5
Groundwater and Subsurface Dissolved Solids Discharge to the Mediterranean Sea

Continents and Islands	Groundwater Discharge			Subsurface Dissolved Solids Discharge		
	Areal Values L/s/1 km²	*Linear Values 100m³/d/km*	*Total Values km³/yr*	*Areal Values T/yr/km²*	*Linear Values 1000 T/yr/km*	*Total Values millions T/yr*
Africa	0.4	3.1	5.1	24.4	2.2	9.9
Asia	2.4	7.0	8.3	110.3	3.6	11.9
Europe	4.0	10.9	33.9	68.4	2.1	18.4
Major Islands	2.8	8.1	5.7	34.9	1.2	2.3
Total			53.0			42.5

favorable conditions for groundwater recharge. Besides, mountain topography, high permeability of fractured solid rocks and terrigenous formations, and a poorly developed river network are responsible for generation of substantial submarine groundwater discharge.

Differentiated estimation of groundwater discharge makes it possible not only to compute values of discharge from separate continents, but also to characterize all the submarine groundwater outflow to individual oceans. Data from Table 4.1 show that the total groundwater discharge to the Atlantic Ocean amounts to 815 km^3 per year, to the Pacific Ocean 1300 km^3 per year, and to the Indian Ocean 220 km^3 per year. The groundwater outflow to the Arctic Ocean has been estimated only for the area of Europe, which does not exceed 50 km^3 per year. On the Asian and American coasts of these oceans, permafrost is developed almost everywhere, which practically excludes generation of submarine groundwater discharge from the upper hydrodynamic zone.

The comparison of values of total groundwater discharge into the world's oceans, calculated by the two methods — hydrogeodynamic (2200 km^3 per year), and described above hydrological-hydrogeologic (2400 km^3 per year) indicates a quite negligible difference. This indicates that the data on groundwater discharge given in Table 4.1 are quite reliable.

Calculations by I.S. Zektser and R.G. Dzhamalov (1988, 1989) of groundwater discharge into the seas and oceans made it possible to quantitatively assess ionic groundwater discharge (surface dissolved solid discharge), to clarify the main regularities of its formation and its contribution to the marine salt balance. Ionic groundwater discharge is a product of common mineralization that is drained by sea aquifers and discharged directly into the sea, bypassing the river system. As indicated above, groundwater discharge directly into seas usually does not exceed some percents relative to river water inflow. However, the contribution of ionic groundwater discharge into the marine salt balance, particularly the interior ones, is considerable enough and amounts to tens of percents relative to salt exportation by rivers. For instance, salt exportation by groundwater in the Caspian Sea is about 27% of that by rivers, while groundwater discharge is a bit more than 1% of the river water runoff.

Ionic groundwater discharge into oceans is caused primarily by submarine discharge, as groundwater mineralization in the intensive water exchange zone usually does not exceed 1 g/l. Analysis of the distribution of calculated groundwater mineralization in all the catchment areas having runoff into the sea indicates that, in most cases, it is about 1 g/l, and only in a few cases does it amount to 4–6 g/l. High groundwater mineralization (15–40 g/l) is characteristic for certain areas of Africa and Australia.

In the distribution of groundwater and subsurface dissolved solids discharge to seas, the general vertical hydrodynamic and hydrochemical zonality for groundwater is observed. This zonality governs the increase in the total transport of salts with depth despite the general reduction in groundwater discharge. This is explained by the substantially larger salinity of groundwater of deep aquifers than the groundwater of upper aquifers. This general law is sometimes disrupted by the effect of local hydrogeological conditions (wide occurrence of karst, presence of salt-bearing rocks, and processes of continental salinization). For instance, the largest values of subsurface dissolved solids discharge to the Baltic Sea (48.5 t per year per sq.km) are characteristic for the coastal portion of the Silurian–Ordovician plateau, where submarine groundwater flow is mainly generated in aquifer systems composed of karstified limestones and dolomites.

Salt exportation by groundwater discharge into the Atlantic Ocean is 479 mln.t per year, into the Pacific Ocean 521 mln.t per year, into the Indian Ocean 296 mln.t per year and into the Arctic Ocean (from estimated catchment areas) 7 mln.t per year. These values indicate that submarine groundwater discharge can significantly affect salt and hydrobiological regime of seas and oceans, and the processes of biogenic sedimentation and mineral-deposit formation. Thus, an essential contribution of groundwater in salt exportation into the world's oceans (52% of salt exportation by rivers) has been ascertained and it radically changes the fully developed idea that initial bioproduction of oceans on the scale of biogenic sedimentation is limited by salt exportation only by the river runoff.

Specific characteristics of groundwater discharge into seas and oceans (areal module and linear discharge) provide the possibility for analysis and comparison of peculiarities in its formation in different physical-geographical and structural-hydrogeological conditions. The relationship between submarine groundwater discharge and the main natural factors forming it is clearest when comparing specific values of groundwater discharge from specific coastal regions of certain continents.

Analysis of submarine discharge conditions for separate continents has been made by R.G. Dzhamalov, who made it possible to reveal main peculiarities in its distribution within every continent in different natural zones of the earth (Dzhamalov et al., 1977; Zektser et al., 1984; Zektser and Dzhamalov, 1988).

In the schematic maps of continents there are both general and specific values of groundwater discharge into seas and oceans, as well as curves of changes in water and ionic groundwater discharge for latitudinal zones (Figure 4.4 to Figure 4.9).

Analysis of conditions forming groundwater discharge into the world's oceans from the continents indicates that this global process depends on a combination of different natural factors, among them climate, relief, and structural-hydrogeological peculiarities of the coastal areas being the most essential. Hydrodynamics of groundwater discharge, filtration, and capacity properties of the vadose zone and water-bearing rocks also considerably affect submarine discharge. All these factors are closely interrelated and cause conditions for groundwater recharge, flow, and discharge in different natural conditions. The highest values of groundwater discharge are characteristic for mountainous coastal regions of tropical and humid zones, where modules amount to 10–15 l/s per sq.km and groundwater discharge in these

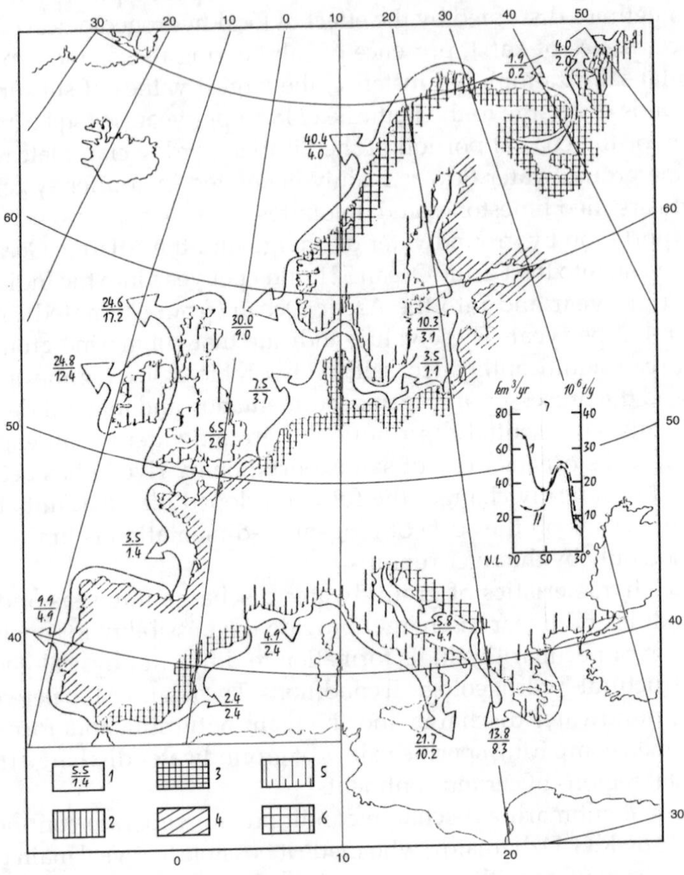

FIGURE 4.4
Schematic map of groundwater discharge to oceans from Europe. (1-the groundwater discharge value in the numerator, km^3/year, the subsurface dissolved solids discharge in the denominator, 10^6 t/year. The groundwater runoff modulus: 1/(sec km^2). 2-0.5–1.0; 3-1.0–2.6; 4-3.0–5.0; 5-5.0–7.0; 6-10.0–15.0; 7-the distribution of groundwater (I) and subsurface dissolved solids (II) discharge to oceans from continent according to latitudes.)

areas amounts to tens of thousands of cubic meters per day for 1 km of the coastal line. The smallest values of submarine groundwater discharge (0.2–0.5 l/s per sq.km) are characteristic for arid and arctic areas in the earth because of unfavorable climate conditions. Groundwater discharge depends on the inflow and outflow contents of water balance in the catchment areas, that are caused by correlation of heat and moisture as the main characteristic of natural physical-geographical zonality. They gradually increase from sub-arctic areas to a moderate zone, abruptly grow in humid subtropic and tropic zones and decrease in semi-arid and arid zones (Figure 4.10). Local oro-graphic, geologic-structural and hydrogeological peculiarities of the coastal catchment areas complicate this general picture of distributing the discharge and sometimes can cause their substantial deviations from mean values char-

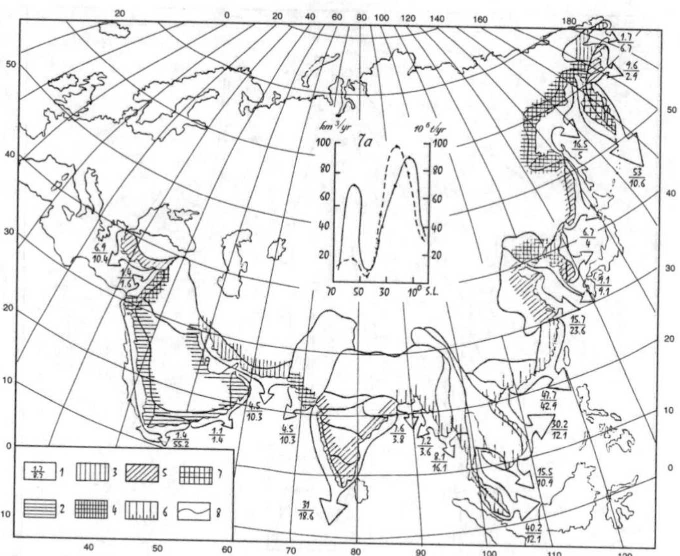

FIGURE 4.5
Schematic map of groundwater discharge to oceans from Asia. (1-the groundwater discharge value in the numerator, km^3/year, the subsurface dissolved solids discharge in the denominator, 10^6 t/year. The groundwater runoff modulus: 1/(sec km^2). 2-0.2–0.4;3-0.5–1.0;4-1.0–2.6;5-3.0–5.0;6-5.0–7.0;7-10.0–15.0; 8-boundary and number of hydrogeological regions and artesian basins; 9-distribution of groundwater (I) and subsurface dissolved solids (II) discharged into oceans from continents according to latitudes.)

acteristic for a given latitudinal zone. Hence, azonally high or low values of submarine discharge connected with the screening effect of mountainous constructions on atmospheric circulation, intensive karst development, the draining effect of river valleys and other local factors, are confined to concrete sites in the coastal zone and, on the whole, do not disturb a general dependence of groundwater discharge into the world's oceans on a latitudinal physical-geographical zonality (Dzhamalov et al., 1977; Zektser et al., 1984).

The comprehensive study of groundwater discharge into the seas and the world's oceans makes it possible to reveal the regularities of this global process, which is inseparably connected with forming, distributing, and transforming the underground constituent of the total water cycle.

4.4 Groundwater Discharge Into Large Lakes

The need to revise lakewater balances and predict changes under intensive human activities in watersheds triggered the study of lake and groundwater interaction. This work was and is being carried out at present in a number of large lakes in the former Soviet Union, in the Great Lakes of North America,

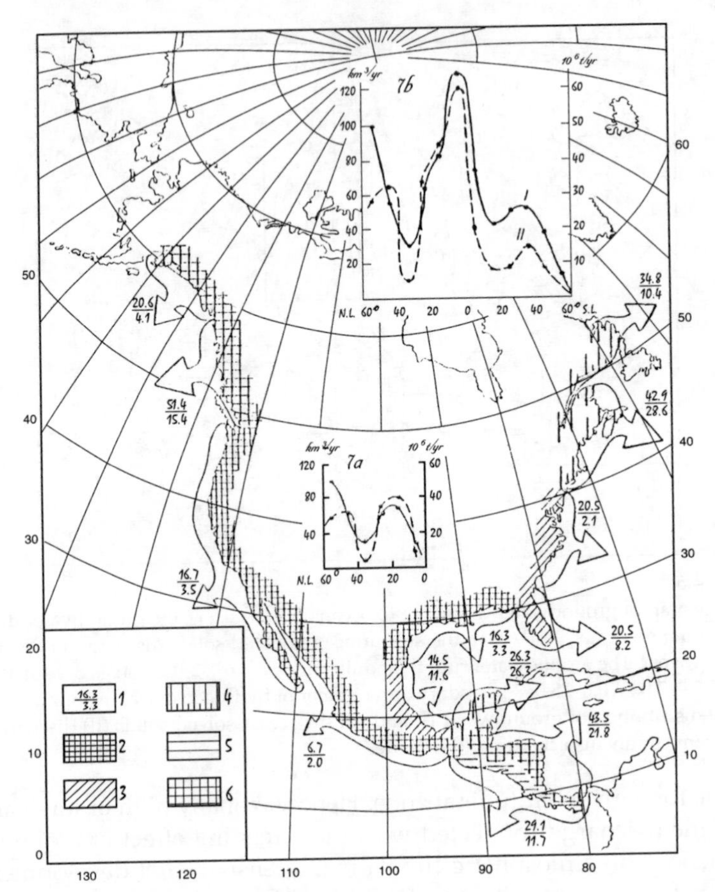

FIGURE 4.6
Schematic map of groundwater discharge to oceans from North America. (1-the groundwater discharge value in the numerator, km^3/year, the subsurface dissolved solids discharge in the denominator, 10^6 t/year. The groundwater runoff modulus: 1/(sec km^2). 2-1.0–2.5;3-3.0–5.0;4-5.0–7.0;5-7.0–10.0;6-10.0–15.0; 7(a)-distribution of groundwater (I) and subsurface dissolved solids (II) discharged into oceans from continents according to latitudes; 7(b)-the same summary from both North and South America.)

and in some other countries. The main results of this work, obtained primarily for lakes in the former USSR, are considered below.

4.4.1 The Caspian Sea

The major problem for the Caspian Sea is significant changes in its water level. For the past 15 years, water level in this interior lake, the largest in the world, has risen by almost 2 m. In recent years, a sea-level decrease occurred that reached 30 cm in 1998. Since lake-level fluctuations cause considerable economic losses, there is a need to determine and substantiate measures for maintaining an optimal water-salt regime in the lake for the national econ-

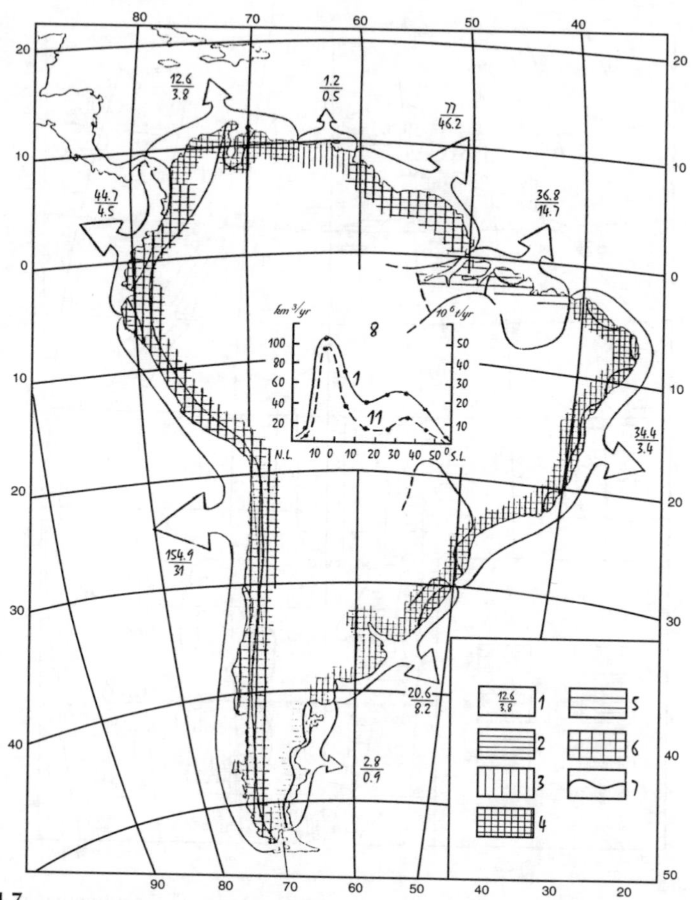

FIGURE 4.7
Schematic map of groundwater discharge to oceans from South America. (1-the groundwater discharge value in the numerator, km^3/year, the subsurface dissolved solids discharge in the denominator, 10^6 t/year. The groundwater runoff modulus: 1/(sec km^2). 2-0.2–0.4;3-0.5–1.0;4-1.0–2.6;5-7.0–10.0;6-10.0–15.0; 7-the boundary of artesian basins and hydrogeological massifs; 8-distribution of groundwater (I) and subsurface dissolved solids (II) discharged into oceans from continents according to latitudes.)

omy. In this connection, the study of modern and historical water and salt balances of the Caspian Sea has become extremely promising.

Numerous assessments of groundwater discharge into the Caspian Sea have been made in previous times. Analysis of these assessments is given by Zektser et al., 1972. However, according to different authors, the volume of groundwater discharged into the Caspian Sea differed by more than 150 times (from 0.3 to 49.3 km^3 per year), due to insufficient reliability of some methods used for the assessment and a lack of reliable hydrogeological data for substantiations. The results of estimating groundwater discharge into the Caspian Sea using a hydrodynamic method (Zektser et al., 1984) are given below.

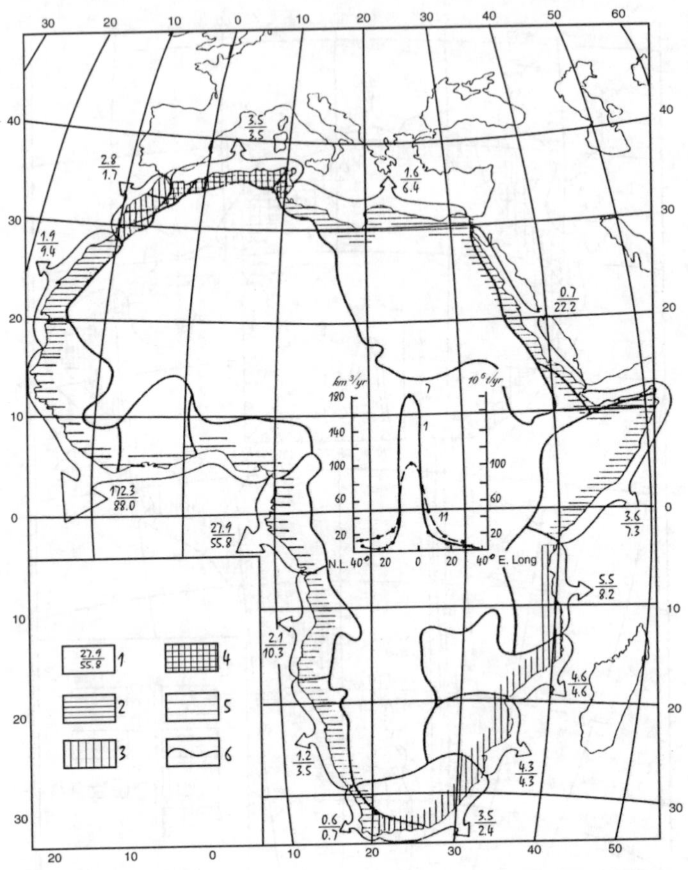

FIGURE 4.8
Schematic map of groundwater discharge to oceans from Africa. (1-the groundwater discharge value in the numerator, km^3/year, the subsurface dissolved solids discharge in the denominator, 10^6 t/year. The groundwater runoff modulus: 1/(sec km^2). 2-0.2–0.4;3-0.5–1.0;4-1.0–2.6;5-7.0–10.0;6-the boundary of artesian basin; 7-distribution of groundwater (I) and subsurface dissolved solids (II) discharged into oceans from continents according to latitudes.)

The Caspian Sea depression, extended latitudinally, is about 1 200 km long and 320 km wide. Its shore length is 7 000 km. There are three depressions in the Caspian Sea basin: the northern one (the Northern Caspian Sea) is a shallow part of the lake with a depth of 5 m and an area of 80 000 km^2, the middle one (the Central Caspian Sea) is an asymmetric basin with an area of 140 000 km^2 and a maximum depth of 788 m (Derbensk hollow), and the southern one (the Southern Caspian Sea) is an asymmetric basin with steep banks and a maximum depth of 1025 m (Kurinsk hollow).

More than 100 rivers flow into the Caspian Sea. The Volga, the Ural and the Terek are the largest and flow into the Northern Caspian Sea, the Volga discharge being 83% of the total river runoff. Mean perennial river runoff into the Caspian Sea is 283 km^3 per year or 790 mm per year.

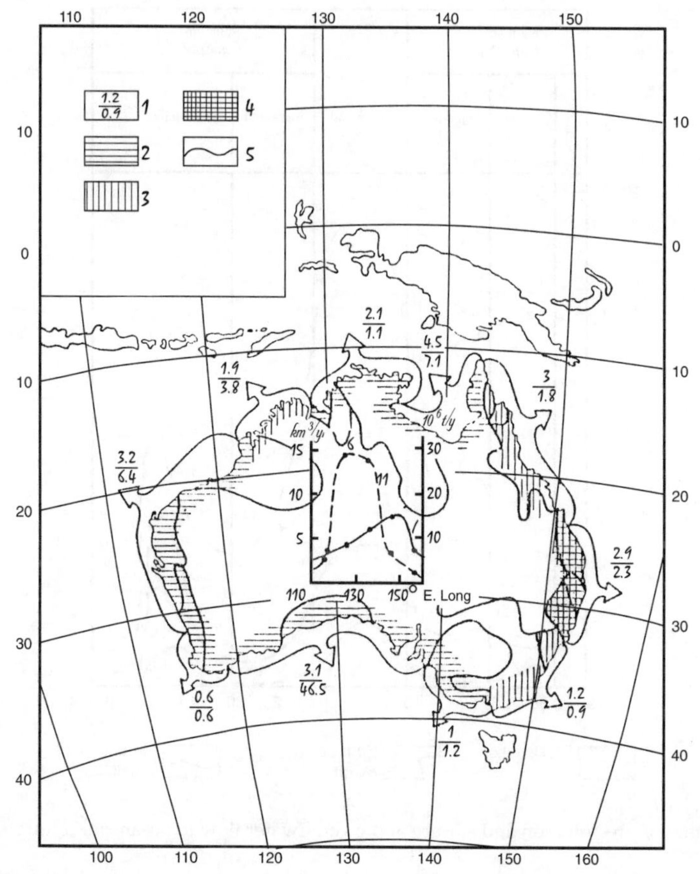

FIGURE 4.9
Schematic map of groundwater discharge to oceans from Australia. (1-the groundwater discharge value in the numerator, km^3/year, the subsurface dissolved solids discharge in the denominator, 10^6 t/year. The groundwater runoff modulus: 1/(sec km^2). 2-0.2–0.4;3-0.5–1.0;4-1.0–2.6;5-the boundary of artesian basins and hydrogeological massifs; 6-distribution of groundwater (I) and subsurface dissolved solids (II) discharged into oceans from continents according to longitudes.)

To calculate groundwater discharge into the Caspian Sea using hydrodynamic methods, data were collected from more than 3800 hydrogeological wells bored in the whole coastal area. Basing the analysis on actual data, the main regional aquifers along the Caspian shore and the main underground component of the seawater balance have been determined. Maps of surface level and water permeability were compiled for these aquifer systems. Groundwater discharge into the sea was then calculated using mainly Darcy's law for groundwater-flow discharge. In this case, calculated hydrogeological parameters were not averaged within considerable areas but taken directly from hydrodynamic maps.

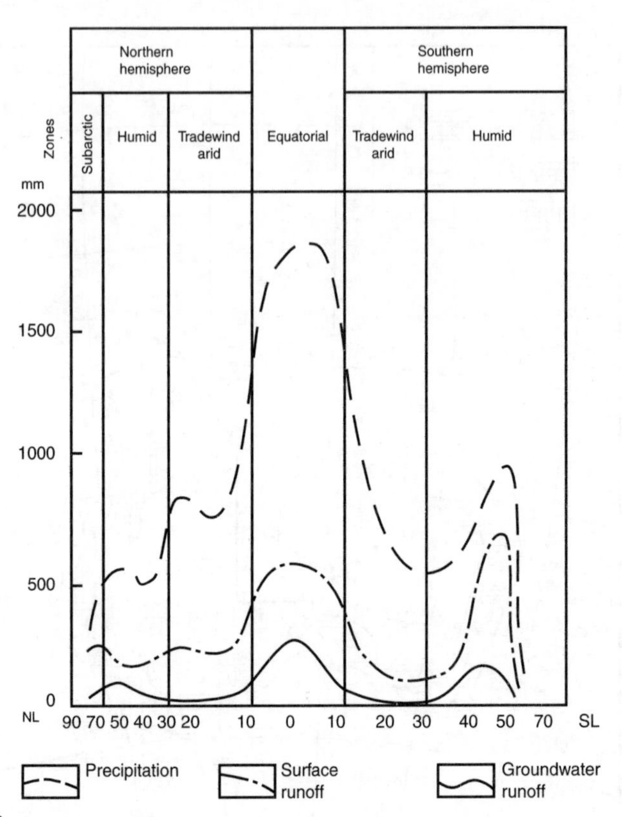

FIGURE 4.10
Distribution of precipitation and surface and groundwater flow to oceans for latitudinal zones of land.

Calculations of flow were made with flow bands covering areas of the coast with hydrogeological conditions and with a width of not more than 30 km. Total groundwater discharge into the lake estimated in this way appeared to be 3.2 km^3 per year. The major factor of groundwater discharge into the Caspian Sea occurs at its western coast due to favorable physical-geographical and geological-hydrogeological conditions (significant precipitation and increased permeability of coarse terrigenous and fractured rocks). Insignificant groundwater discharge occurs on the eastern coast primarily because of a dry climate, low filtration properties of the water-bearing rocks in the coast, and an insignificant gradient of the piezometric water table.

The assessment of subsurface dissolved-solid discharge was made for the whole perimeter of the Caspian Sea, which was subdivided into 15 sections with similar hydrogeological conditions and calculated parameters, and almost similar mean groundwater salinity in the main aquifers discharging into the sea. A total volume of subsurface dissolved solids discharged into the Caspian Sea from the coast amounts to 23 mln.t per year. Thus, even though the volume of groundwater discharged into the Caspian Sea is insignificant and just exceeds 1% of the surface water inflow, the subsurface dissolved solids discharge amounts to about 27% of that from the surface sources and is extremely important in the formation of hydrochemical and hydrobiological regimes (Dzhamalov et al., 1977; Zektser et al., 1972, 1984).

4.4.2 Lake Baikal

Lake Baikal is one of the largest and deepest continental freshwater lakes on the earth. Its area is 31.5 000 km^2, mean depth is 730 m, maximum depth is 1,620 m, and catchment area is 557 000 km^2.

The Lake of Baikal depression is a deep fault, formed by geological shear. From a hydrogeological point of view, it is an artesian basin limited to peaks on the mountain ridges.

The problem of underground water exchange in Lake Baikal was noted in the 1950s in connection with calculation of mean perennial water balance for the lake. In the first stages of the investigation, country data were obtained for the groundwater discharge. Thus, Tseitlin (1959), having calculated the lake's water balance for the period from 1899 to 1955, determined that the outflow component of 1.17 km^3 per year exceeded the inflow component. He considered that the difference was due to outflow from the lake. Later, Vikhulina and Kashinova (1973) found no disparity in the lake's water balance for the period from 1970 to 1971, and considered that there was no underground inflow to the lake. But finally, Afanasiev et al. (1976), having reassessed a perennial balance, found that the disparity (2.3 km^3 per year) corresponded to the underground inflow into the lake. The differences in calculations were the result of errors in calculating other components of the balance. Later attempts were made to estimate groundwater runoff into the lake using a hydrodynamic method (Afanasiev et al., 1976; Pissarsky, 1973). In 1980–1981, the Baikal coastline was surveyed by aerial photography and an examination was made of the coastal zone and water surface, as well as hydrometric measurements of base flow. Infiltration parameters of the water-bearing loose sediments and crystalline rocks were calculated based on pumping test data and observations of the groundwater level regime. The analysis of the data showed that underground runoff into the lake included underground runoff in the river valleys, groundwater inflow from the interstream areas bypassing the river network, spring runoff from the coastal area, and subaqueous inflow directly into the lake. A brief description of these generic components of the underground inflow into Lake Baikal is given below.

An underground runoff into the lake was calculated using the hydrogeological parameters from wells bored in loose sediments of the large rivers (the Selenga, the Upper Angara) and some rivers of the eastern coast of southern Baikal. The underflow runoff was estimated in two ways:

1. by calculating the underflow discharge using hydrogeological parameters
2. by calculating the losses of a part of the river runoff for infiltration in the coastal zone

The method of analogies was used for some areas (e.g. the eastern coast of southern Baikal) where there was a deficiency in hydrogeological data. On the whole, the present studies indicate that the underflow runoff into Lake Baikal, estimated by the method described, is 1 km^3 per year, and is about 1.5% of total river runoff into the lake.

According to periodic observations, a spring runoff into Lake Baikal is estimated as 63 mln.m^3 per year. It includes cold and thermal springs as well as groundwater discharge in the lowland coast of northern Baikal in the form of effluent seepage and water holes. It should be noted that the temperature of six thermal springs discharging in the northern Baikal coast with a total yield equaling 1.6 mln.m^3 per year was 80°C, and so it affected geothermal and microbiological regimes of the lake water in the coastal area.

Groundwater discharge to the lake territories, which bypasses the river network, is very small due to the extremely low filtration properties of the crystalline rocks along the slopes.

Subaqueous groundwater discharge directly into the lake basin is confined to the active fault zones on the Baikal bottom. Both cold fresh and highly saline thermal water are discharged to the bottom of Baikal. Some temperature anomalies in the bottom water, exceeding the background values by 0.5°C, are caused by thermal groundwater discharge at considerable depths. Anomalous concentrations of sulfate, chloride, and sodium, exceeding background concentration by 2–4 times (Misandrontsev, 1975), were discovered in some bed-load deposits at depths from 200 to 900 m. These anomalies can be caused by discharge of strongly thermal and highly saline groundwater. Subaqueous runoff into Lake Baikal roughly amounts to 5 mln.m^3 per year.

The total groundwater discharge into Lake Baikal from all generic components is estimated to be 1.1 km^3 per year, that is, 2% of the total water inflow into the lake.

4.4.3 Lake Balkhash

Lake Balkhash is the largest moderately saline lake in Kazakhstan. Its surface area is 16.4 000 km^2, its length is 600 km, and its width is from 5–71 km. Non-uniform salinity in the lake is a characteristic peculiarity. The water in western Balkhash (the area is 9.6 000 km^2) is relatively fresh, the salinity being 1–2 g/l, whereas water in the eastern Balkhash (an area of 6.9 000 km^2) is moderately saline, the salinity being 3–5 g/l. A decline in the lake's water level has recently been observed due to cyclic fluctuations of the level along with intensive human actives in the catchment area — especially the contribution of the Kapchegeisk reservoir.

The Lake Balkhash basin accumulates the runoff from surface and groundwater of a large catchment basin with an area of 400 000 km^2.

An exchange of underground water between the land and the lake is a very complex process that occurs under highly arid conditions and greatly affects the salt balance of the lake.

TABLE 4.6
Ion Groundwater Discharge Into the Lake of Balkhash

Flows	Groundwater discharge	Mean dissolved solids content in the groundwater (g/l)	Contribution of dissolved solids (t/yr)	Specific groundwater discharge (1/s per km)	Specific flushing of dissolved solids (g/l per km)
Western Balkhash					
The Ili river delta	0.8	1.8	1410	0.09	1.15
Fissure water of the Northwestern Balkhash	5.2	2.0	10400	0.54	1.08
Fault zones	1.01	1.0	1000		
Total	7.0	1.0	12800		
Eastern Balkhash					
The Karatak, Aksy, Lepsi river valleys	2.3	0.8	1910	0.23	0.19
Fissure water	4.3	2.0	8600	0.43	0.86
Fault zones	2.2	1.0	2200		
Total	8.8	12700			
Total for the Balkhash	15.8	25000			

TABLE 4.7
The Main Characteristics of Groundwater Flows in the Balkhash Lake

Groups of flows	Width of flows (km)	Thickness of an aquifer (m)	Transmissibility (m^2/day)	Flow discharge (10^6 m^3/yr)	Specific discharge for 1 km of coast (1/s)
Ili	32	10-50	60-119	0.8	0.8
Karatal	30	30-70	80-190	2.1	2.2
Alsy-Lepsinskaya	6	10-40	40-150	0.2	0.5-1.7

Unconfined groundwater runoff directly into the lake occurs mainly in the coastal zone, 30–35 km wide, in the form of groundwater flows with a different discharge. Groundwater flows are confined to ancient and modern river valleys and large tectonic fault zones. Each group of flows has its own hydrogeological characteristics that cause their discharge. On the whole, there are 10 groundwater flows in the coastal zone of Lake Balkhash. The main parameters of the largest ones and their discharge are given in tables 4.6 and 4.7.

Calculations of groundwater runoff into the lake were made using a hydrodynamic network of flows constructed on the basis of mean perennial groundwater level values. On the whole, groundwater discharge into the lake is estimated as 15.8 mln.m^3 per year and specific flow rates change from 0.5 to 2.5 l/sec per 1 km of the coast in its different parts. Different amounts of runoff occur along the coast: This runoff is 31 mln.m^3 per year in the south, and 127 mln.m^3 per year in the north and west. These differences are due to different filtration properties of the water-bearing rocks.

Groundwater level depression in the lake vicinity is an important peculiarity of groundwater exchange in the lake. The greatest depression is in the ancient delta of the Ili River, 30 km from the coast. Groundwater levels within this depression are 8–9 m lower than the level of the lake and its salinity is 7–10 g/l. Simulation of paleohydrogeological conditions of the Ili River delta showed that the formation of such a depression in the groundwater level is due to perennial water-level fluctuations in the lake.

Assessment of the substances dissolved in the groundwater discharging into the lake was made on the basis of groundwater discharge linear values and data on dissolved solids content. This calculation showed that salt mass in the groundwater runoff into the lake is 25.5 000 t per year (Table 4.6).

It should be noted that a backward gradient of the groundwater flow, leading to an outflow of the lake water into the coast, formed during the phase of the lake transgression lasting 20–25 years when its level exceeded 341 m in the coastal zone. Groundwater discharge out of the lake in these periods is estimated to be 28 mln.m^3 per year, and, within the southern coast only, this discharge exceeds the total groundwater discharge into the lake by 1.8 times.

The situation with a mass of ions is analogous. During the lake transgression, the salt flushed into the groundwater amounted to 42 mln.t per year. This basic difference in underground exchange between water and salt in the lake during its transgression and regressions led to a formation of different hydrochemical regimes at different stages.

The assessment of a confined groundwater discharge into the lake was made based on calculations of vertical groundwater leakage through poorly permeable deposits on the lake bottom under a hydrostatic head. A total groundwater discharge into the lake bottom is 0.06 km^3 per year under a mean module of 0.11 l/sec per 1 km^2. Dissolved solids in the groundwater are slightly more than 1 mln.t per year with an average salinity of 18 g/l.

Thus, about 0.08 km^3 per year of groundwater (confined and unconfined) discharges directly into the lake, and dissolved solids amount to 1.1 mln.t per year. This is about 0.5% of the surface water inflow and about 26% of the salts brought with it.

Hydrogeologists in Kazakhstan made a tentative prediction of changes of the groundwater discharge into the lake with a further decline of its level. Calculations were made using mathematical modeling of profiles when the water level of the lake has fallen by 1, 2, 3, and 4 m. It was shown that underground, surface, and salt runoff into the lake basin will increase slightly, and direct discharge into the lake will not actually change.

4.4.4 Lake Issyk-Kul

Lake Issyk-Kul is situated in the southern part of the former USSR near the northern Tien-Shan mountains and is the largest water body in the mountainous area. Due to population growth and increasing economic development of the lake basin, it was necessary to scientifically substantiate measures for rational use and protecting natural resources of the lake and adjacent areas.

The study of the hydrological and salt balances of the lake was an integral part of these investigations (e.g., Kadyrov, 1986).

Lake Issyk-Kul is located in the center of an intermontane area of the same name. An artesian basin underlies this area. It consists of three hydrogeological strata:

1. an upper hydrogeological stratum composed of loose Quaternary sediments
2. a middle hydrogeological stratum of consolidated Mesozoic-Cenozoic rocks where substantial supplies of confined groundwater are stored in sandstone and gritstone interlayers alternating with clay layers
3. a lower hydrogeological stratum composed of semipermeable rocks that does not significantly contribute to groundwater discharge

Some important water areas of Lake Issyk-Kul's basin were selected for detailed investigations of a groundwater-discharge process. Seepage meters were placed at 69 points to measure a rate of groundwater seepage into the lake at the bottom–water interface. Vertical probing of bottom sediments for temperature and conductivity was made at 16 points. From inherent relationships, the distribution of salinity in the groundwater percolating though the bottom sediment was calculated. A vertical distribution of head in the ground water saturated bottom sediments was studied at 9 points (Meskheteli et al.,1987). Samples of discharging groundwater were taken from seepage meters for chemical and isotopic analysis. Horizontal temperature and conductivity profiling was also carried out along 22 lines.

The results obtained allowed the preparation of detailed schemes of intensive groundwater discharge into water areas. Hydrogeological investigations in the water area of Lake Issyk-Kul made it possible to determine the value of groundwater discharge to the lake from all the aquifers. The rate of groundwater discharge from the upper hydrogeological stratum to the lake was evaluated by a hydrodynamic method.

A groundwater discharge from the middle hydrogeological stratum for each region was estimated as the difference between the groundwater discharge value obtained experimentally and the results of hydrodynamic computations for the upper hydrogeological stratum. Measurements made on land and in the lake allowed the chemical composition of groundwater discharging within each area to be determined.

The data on subsurface water and dissolved salt discharge to Lake Issyk-Kul are presented in tables 4.8 and 4.9.

According to these estimates, the groundwater discharge to Lake Issyk-Kul amounts to 1.5 km^3 per year. This groundwater carries 755,000 t per year of dissolved salts into the lake. As for the contribution of groundwater outflow to the salt composition of Lake Issyk-Kul based on these estimates, groundwater delivers 60% of the total supply of chloride, 62% of sulfate, 44% of bicarbonate, 70% of sodium and potassium, 49% of calcium and 79% of

TABLE 4.8

Transport of Dissolved Salts by Groundwater to the Issyk-Kul Lake

		Dissolved Solids Discharge (kg/s)										
Region	G.D. m^3/s	Total	Cl^-	SO_4^{2-}	HCO_3^-	$(Na+K)^+$	Ca^{2+}	Mg^{2+}	NH_4^+	NO^{3-}	NO_2^-	H_4SiO
I	31.7	11.4	0.59	1.57	5.84	0.77	1.89	0.49	0.01	0.02	0.0003	0.40
II	8.1	2.9	0.14	0.40	0.73	0.13	0.25	0.09	0.001	0.006	0.0001	0.03
III	5.9	4.2	01.87	1.1	2.68	0.94	1.40	1.54	0.002	0.054	0.0002	1.62
IV	1.6	0.9	0.028	0.026	0.17	0.057	0.05	0.01	0	0.005	0	0.02
Total	47.3	19.4	1.63	3.13	9.42	1.89	3.59	2.13	0.013	0.85	0.0005	2.07

G.D. = groundwater discharge (47.3 m^3/s = 1.5 km^3/yr)

TABLE 4.9

Supply of Minor Elements by Groundwater to the Issyk-Kul Lake

	Dissolved Solids (ions) Discharge (g/s)												
Region	Fe	Cu	Zn	Pb	As	Mo	Ni	Co	I	Br	F	Mn	P
I	3.8	0.36	0.78	0.028	0	0.014	0.005	0	0.19	0.19	5.8	0.034	0.35
II	0.21	4.5	13	0	-	-	-	-	-	-	0.39	-	0.24
III	0.90	0.031	0.065		0	0.023	0	0	0	0	5.9	0.04	0.10
Total													
(g/s)	4.9	4.8	13.9	0.028	0	5.4	0.05	0	0.19	0.19	11.2	0.074	0.69
(t/yr)	155	151	438	0.88	0	170	170	-	6	6	350	350	22

magnesium. A great quantity of minor elements is also discharged into Lake Issyk-Kul (Table 4.8).

It should be noted that pollutants can also be easily transported by groundwater discharge to the lake. This is because groundwater areas are composed of coarse-pebbled sediments, which facilitates penetration of polluted surface water into the aquifers. High rates of groundwater flow may then lead to a rapid transfer of pollutants into the lake. In the Issyk-Kul catchment area, there are many small factories and farms from which pollutant discharge is possible. The diffuse nature of pollution sources in the Lake Issyk-Kul catchment complicates groundwater quality monitoring and groundwater pollution control, which may have a negative effect on the water quality of the lake. An example is the high supply of nitrogen-containing compounds to the lake (Table 4.8). Estimates of the pollutants delivered by groundwater to the lake are given in Table 4.10.

Thus, the results show that despite a relatively small amount of groundwater inflow into the lake, its effect on the salt composition and quality of lake water is highly significant, and comparable to that of surface water. On the whole, groundwater inflow accounts for 30–40% of the total water inflow into the lake (river runoff, groundwater discharge, and precipitation) and transports over 50% of the total dissolved solids (Bergelson et al., 1986).

4.4.5 Other Lakes

Investigations of surface and groundwater interaction are not confined to lakes in the former Soviet Union countries. Studies of groundwater discharge into Lake Michigan, for example, are made in the Center for Studying the Great Lakes in the USA, and an assessment of groundwater inflow into Lake Michigan from coastal areas has been made. Using a hydrodynamic method for calculating groundwater discharge, the groundwater inflow into Lake Michigan in the studied territory and under natural conditions has been determined to amount to 580–880 m^3/day per 1 km of the coast line. The amount of groundwater discharge of the total water inflow into the lake in this sector is estimated to be 7–11% compared with surface water inflow. Human activities, particularly intensive groundwater pumping, decreases the water inflow into the lake.

TABLE 4.10
Approximate Estimates of Some Pollutants Delivery by Groundwater to Issyk-Kul Lake

	Pollutants Delivery Rate (g/s)			
Region	Phenol	Petroleum products	Synthetic surface active substances	Agricultural chemicals
I	0.053	0.16	0.28	0.001
II	0.003	0.15	0.022	0.002
III	0.009	0.29	0.011	0.016
IV	-	-	-	-
Total (g/s)	0.065	0.60	0.313	0.19
(t/yr)	2	18	9.9	0.6

5

The Environment and Groundwater Pollution

5.1 The Interconnection Between the Pollution of Groundwater and the Environment

The use of groundwater for water supply is limited not only by its quantity, but also by its quality.

The idea that groundwater is ecologically pure should be considerably corrected in some regions. Groundwater pollution is mainly of a local character, however, polluted areas amount to tens and in some cases even hundreds of square kilometers.

In one book and especially in a single chapter that is limited in volume, it is impossible even to list the examples of interconnection between the pollution of groundwater and the environment. Some particular examples are given in the next section, where recent conditions of groundwater pollution in Russia are described. Here, it must be mentioned that the content of oil products, phenols, nitrates, and other components in the groundwater in some places exceeds maximum permissible concentrations (MPC) by tens and even hundreds of times. Regional groundwater pollution is often caused by acid rain and nuclear-power stations. In some regions, acid rain and melted-snow infiltration have increased the aggressiveness of groundwater and have caused regional changes in the degree of oxidation reduction (redox) conditions and an increase in sulfate and heavy-metal concentrations.

Pollution can be the primary reason for a water crisis. Only about 5% of industrial and domestic wastes in towns of developing countries are subjected to any type of treatment. The rest, including two million tons of human excrement daily, as well as toxic and dangerous industrial by-products, are carelessly disposed of and pollute soil, rivers, and aquifers.

In most cases, groundwater pollution is an immediate result of environment pollution. Actually, any human interference in nature and any kind of human activities (hydrotechnic and civil engineering, mineral-deposit mining, forest chipping off, addition of fertilizers to soil, etc.) inevitably affect groundwater quality and resources. Therefore, groundwater pollution — especially groundwater protection — are closely connected with environmental protection.

Various human activities bring about considerable changes in the conditions of groundwater-resources formation, causing their depletion and pollution in many cases. Toxic-waste disposal into deep aquifers is also a serious danger. Groundwater is mainly polluted by sulfates, chlorides, nitrogen compounds (nitrates, ammonia, ammonium), oil products, phenols, iron compounds, and heavy metals (copper, zinc, lead, cadmium, mercury).

For instance, intensive anthropogenic impact on nature in southern California, resulting from oil fields development, brine disposal into aquifers, intensive use of pesticides and other fertilizers has caused an increased arsenic content (up to 0.03–0.04 mg/l) in the groundwater.

Most dangerous, and unfortunately, the most widely spread type of environmental pollution is oil pollution. Oil pollution of the environment manifests itself not only near areas of oil production and refining, but also near oil storage facilities and pipelines. According to the data of American specialists, about 50% of oil storage sites in the United States are leaking. Considerable leakage into groundwater occurs in almost all of the numerous gas station facilities.

Leakage of oil and chemicals from refineries, oil tanks and pipelines are the most hazardous. They penetrate through the soil into the water-bearing layers, causing groundwater contamination. Oil products in the underground strata also give off fumes that migrate through soil and endanger the health and even the lives of humans.

The situation in the Volga River area can serve as an example. The largest oil refineries in Russia are located there. These plants were built in the 50s, when all pipelines were laid at a depth of 5 m. This did not effectively control oil, and now oil products are leaking from the pipelines. At that time, the largest power electric stations were built on the Volga. In particular, the filling of the Saratov water reservoir and the subsequent water level rise of up to 5–6 m in the Volga, caused flooding of the flood-plain and a groundwater-level rise that caused groundwater and soil contamination. Oil products, poisonous to people, appeared in water wells. In the spring of 1989, a thick layer of oil products emerged in the cellars of apartment houses at levels close to the high flood-plain level in the town of Novokuibyshevsk, in the Samara region. Fires and cases of poisoning appeared. The situation could be called an ecological catastrophe. In the spring of 1991, there were more victims of this catastrophe. A federal commission investigated the reasons for the tragedy and found that a thick layer of oil products had formed at a depth of 3 to 70 m. In the town of Novokuibyshevsk alone, the layer of oil products, mainly benzene, amounted to 1.1 million tons. The oil products migrated with the groundwater toward the river and contaminated the river water. A complex set of geological studies, gas surveys, seismic and electric exploration were carried out in the area. This made it possible to develop a model of oil-product migration and to confirm the measures needed for the purification of the soil and groundwater. These measures included biological purification with microorganisms, cleaning tanks, and settling basins in the oil refineries, pumping of oil products (with their reuse) from anthropogenic layers, wastewater treatment, and degassing, etc.

Oil pollution often does not manifest itself until it has already reached a catastrophic level, which makes it even more dangerous. Experience obtained in different CIS countries (Russia, Ukraine, Kazakh Republic, and the Uzbek Republic) indicates that almost all the activities of the oil industry are, to a certain degree, sources of environmental pollution (Borevsky et al., 1994).

World experience has shown that the results of the purification of the environment from oil-product pollution do not compare with the scale of measures and financial investments in the clean-up. The work carried out in the USA "Superfund" is particularly demonstrative, as there was no situation where the aim of complete environmental rehabilitation was achieved in spite of the expenditure of many billions of dollars. This is because the purpose of this work — to entirely purify the environment of already formed, stable pollution — is unreal.

The analysis of experience in environmental rehabilitation (primarily geological) in about a hundred cases indicates that rehabilitation should not be aimed at bringing the pollutant content in water and soil into line with MPC, but at protecting water usage (potable well fields, surface reservoirs and channels). So, pollution must not exceed a level that is safe for the population and the environment. Thus, the strategy of coping with oil-product pollution of aquifers must be aimed at protection from pollution (Borevsky et al., 1996).

Various waste dumps are a source of long-term negative effects on the environment. For instance, there are about 40 000 dumps in Germany, and 30 000 in Russia. There are more than 45 billion tons of tailings in the mining industry. At present, there are 150 old and 100 new dumps in Moscow. Waste dumps are often in unacceptable hydrogeological conditions (absence of confining beds, protecting aquifers). They pollute fresh groundwater and, in some cases, considerably limit its use for potable-water supply.

Groundwater protection from pollution should incorporate:

1. common measures aimed at environmental protection on the whole:
   - the realization of technical and technological measures for decreasing by-products
   - creating waste-free by-products
   - water reuse
   - preventing wastewater leakages
   - controlled and limited use of toxic chemicals and fertilizers
2. special measures:
   - organization of aquifer sanitary-protection zones
   - the revealing of actual and predicted potential pollution sources
   - the adoption of protective measures for liquidation and localization of available pollution sources and prevention of their forming in future
   - careful choice of sites for building new industrial constructions and agricultural objects
   - detailed hydrogeological proving of underground industrial-waste disposal, etc.

Development of complex monitoring systems that incorporate atmospheric precipitation, surface water, the vadose zone, and groundwater is particularly important. Stationary observations at potential places of water pollution must be a part of this monitoring in industry and agriculture.

The purpose of monitoring water quality in the aquifer and vadose zone is the prevention of possible pollution. One of the prediction methods is the construction of permanent analogues (PA) of aquifer systems using computers.

World experience shows that it is impossible to work out water-protection policies under market economy without the support of corresponding public opinion. Scientists warned long ago about the catastrophic state of some of the largest water projects. However, governments started with their programs (e.g., the Great Lakes, the Rhine, the Thames) only after the ecological situation had become critical and every ordinary person had become aware of the catastrophe.

5.2 Human Impact on the Groundwater

Analysis of groundwater resources and water-quality changes affected by intensive anthropogenic activities is of great practical value for assessing and predicting groundwater's possible use as a primary source of domestic and potable-water supply (Borevsky, Yazvin, 1991; Yazvin, Zektser, 1996b).

These activities include:

- groundwater pumping for use as a public and industrial water supply, for land irrigation and drainage, and pasture watering
- development of solid mineral resources, oil and gas fields
- industrial and civil engineering, operation of engineering structures
- agriculture, including land irrigation and drainage, as well as forest-industry activities
- hydraulic construction, and construction and operation of nuclear-power plants

These human activities result in variations the recharge, storage, and safe yield of groundwater. These variations generally cause a disturbance of groundwater recharge and discharge conditions (ratios between inflow and outflow components of the groundwater budget) and deterioration of groundwater quality due to anthropogenic pollutants. This constitutes encroachment of substandard natural water into aquifers or surface water bodies.

The impact of the anthropogenic activities given above on the resources and fresh groundwater possible use will be briefly considered here.

Groundwater Withdrawal

The most substantial changes in the natural groundwater balance and, therefore, in groundwater resources, are brought about by heavy groundwater withdrawal for public and industrial water supply. The withdrawal carried out by means of numerous well fields and individual wells results in formation of cones of depression, changes in groundwater-flow direction, and transformation of discharge areas into recharge areas.

Groundwater withdrawal influences resources differently under different hydrogeological conditions. Development sometimes results in an increase of groundwater recharge, which is due to a change in evaporation from the water table (i.e., groundwater discharge into the vadose zone).

As I. Pashkovsky (Kats and Pashkovsky, 1988) indicated, because of a decrease in evaporation from the water table, the resultant groundwater recharge increases down to certain depths and becomes constant in the zone where the evaporation may be neglected. Evaporation from the water table depends on a number of factors, among which the depth to the water table (the thickness of the vadose zone), the lithological composition of the rocks of the vadose zone, and climatic conditions, are principal. In summer, where there is a shallow water table, groundwater evaporation may be very substantial even in a humid environment. The resultant groundwater recharge may increase appreciably with even a small decline of the water table.

In arid zones, a groundwater-level decline from 1 to 2 m can increase the resulting recharge rate by 100–2 mm per year (Plotnikov, 1989). Calculations of the groundwater balance for some key sites in humid zones, particularly in Latvia (Sakalauskene, 1977), indicate that at the depth where evaporation from the groundwater surface ends, the resulting recharge can be almost doubled. Groundwater-level decline due to well-field operations causes an increase in the resulting groundwater recharge, which affects its safe yield formation and volume. Thus, according to D.I. Efremov's data, in the northern part of the Moscow region, insignificant groundwater-level decline causes an increase in recharge. This can be 60-70% of the total groundwater safe yield in Carboniferous aquifers, which are widely used for water supply.

Groundwater pumping by large well fields generally leads to pollution of the water being withdrawn. However, groundwater pumping sometimes improves water quality. An iron-free groundwater lens formed as a result of long-term withdrawal on the Zayachii site of the Ostrovnoe groundwater reservoir in the alluvial deposits of the Amur River; this water lens originated as a consequence of the Amur water encroachment (Kulakov, 1990).

Development of Mineral Deposits

Development and exploitation of solid mineral deposits is frequently accompanied by drainage and drawdown operations. These operations produce effects similar to those of groundwater withdrawal — variations in groundwater recharge and discharge conditions, appearance of large cones of depression, aquifer depletion, and changes in groundwater chemical composition. Unlike groundwater development for water supply, the groundwater withdrawal in mining areas is taken from all the producing strata. At present, the pumping depth may reach 600 m, which results in greater drawdowns that in groundwater withdrawal by water supply wells.

If the roof caves in where mineral deposits are developed, the permeability of the overlying rocks improves and water-conducting fractures form. These conditions often lead to an increase in infiltration and recharge of groundwater. Construction of tailing ponds and water disposal sites is an important factor that brings about a change in groundwater resources in the course of mineral-deposit development. These structures are responsible for increased water inflow to mines and quarries and cause groundwater pollution.

When developing solid mineral deposits, a specific chemical composition of groundwater is frequently formed that results from the mixing of water from various aquifers, interaction with enclosing rocks, and pollution in mine workings. In Russia, the development of coal deposits in the Kuznetsk, Moscow, Kizel, and Pechora basins, the iron-ore deposits of the Kursk Magnetic Anomaly area, the bauxite deposits of the northern Urals region, and other areas have the largest effect on hydrogeological conditions and groundwater resources.

Development of solid mineral deposits results in the depletion of a safe groundwater yield. This also involves discharge of pumped water in the area of the mineral deposit, failure of operating water supply wells, groundwater-level decline in water-promising areas, and groundwater pollution. Based on the analysis of available data, the largest cones of depression are formed when regional aquifers discharge their water into mine workings.

Pollution of groundwater in solid mineral deposits depends, in many respects, on the methods of drainage-water control in mines and quarries. The chemical composition and total-dissolved-solids content of the groundwater beyond and within the mine workings differ considerably. The groundwater of mine workings is oxidized as a result of intensification of rock leaching, changes in the gas and bacterial composition, and penetration of petroleum products, oil, and suspended matter. Acid groundwater often forms in coal deposits, and higher concentrations of microelements (Cd, Zn, Cr, Sr, Ni, and others) are observed in the water in ore and coal deposits. Higher contents of sulfates, chlorides, and increased hardness are frequently recorded.

Thus, the main hydrogeological problems in exploiting solid mineral deposits are groundwater pollution and depletion control to maintain safe yield. The solution for these problems requires a new, comprehensive approach to constructing drainage systems (Yazvin, 1992). A modern, efficient system of protecting a mineral deposit from drainage water should produce the required drainage effect. However, parallel mine drainage and use of groundwater for the water supply of facilities in the area of mineral deposits places new requirements on the drainage method and the location of water supply wells. Maximum possible measures for groundwater-quality preservation should be specified, with simultaneous efficient drainage. For this reason, external drawdown systems located at an optimal (favorable for combined drainage-water supply operations) distance from boundaries of a mine or quarry should be preferred. The use of drainage water for production and public water supply demands special hydrogeological investigations during exploration and development of mineral deposits. The goal of these investigations is the assessment of the drainage water safe yield, which is considered to accompany the mineral resource. Methods for safe-yield assessment and special features of hydrogeological investigations of mineral deposits (explored and developed) are discussed in detail in the methodological recommendations published by the All-Russia Research Institute for Hydrogeology and Engineering Geology (Methodological Recommendations ..., 1992).

Groundwater is injected into oil- and gas-bearing productive beds to maintain reservoir formation pressure. This may result in pollution of both fresh and brackish water of shallow aquifers by oil and highly mineralized water filtration along fractured zones. Oil leakage from an oil-bearing formation along a fractured zone was observed in one of the oil fields in the Perm region, where the nature of water pollution in the Kama Reservoir by oil was revealed (Oborin et al., 1994). Injection of large quantities of fresh surface water in the course of peripheral and contour flooding may lead to dilution of valuable mineral waters and brines. It should be emphasized that the processes causing variations in groundwater resources and quality under development of oil and gas fields have been studied insufficiently.

Industrial and Civil Engineering and Operation of Engineering Structures

Industrial and civil activities influence groundwater quality as well as recharge and discharge conditions differently. In some cases, human impact results in an increase in groundwater recharge caused by:

- leakage from water pipelines and sewage ponds
- by infiltration of wastewater and irrigation water
- moisture condensation under structures and asphalt pavements
- seepage from ponds and reservoirs
- backwater formed by embankments and deep foundations

In other cases, human impact causes an increase of groundwater discharge, i.e., groundwater development, outflow from foundation pits, drainage on subway lines, and various drainage operations, or a decrease in groundwater recharge, e.g., asphalt pavement and snow removal. An increase in groundwater recharge from leakage of industrial wastewater, in particular, may cause appreciable negative changes in groundwater quality.

An increase in groundwater recharge, i.e., generation of artificial groundwater resources and artificial groundwater storage, is accompanied by a rise in groundwater levels that can result in flooding of urban areas. Thus, a new anthropogenic aquifer was formed over 20 years in the area of Nizhni Nivgorod agglomeration. The area of this aquifer quadrupled in 15 years and the level rise amounted to 10 m (Dzhamalov and Grishina, 1987). Seepage losses from reservoirs, ponds, and water pipelines, as well as watering of cultivated vegetation, resulted in a threefold increase in groundwater recharge in the city of Moscow compared with undeveloped areas (Borevsky et al., 1989).

Control of groundwater level rise requires special drainage measures using groundwater pumping to reduce the water level to a normative value. In some cases, this drainage effect can be achieved by an operating well field that supplies people with water. Here groundwater well fields provide a double function: They supply the population with quality water and improve conditions for building and functioning of underground constructions, such as a subway, by limiting water inflows to them. Therefore, ceasing well-field use can cause negative consequences for underground constructions. For example, stopping groundwater withdrawal for water supply in Brooklyn, NY has caused the groundwater level to rise and foundations and subway tunnels have subsequently been flooded.

Seepage from storage facilities plays the most negative role in groundwater pollution. These facilities include:

- sewage, tailing, storage, evaporation, retention, and slime ponds
- oil and petroleum product storage facilities
- storage facilities for solid industrial wastes (ash dumps and salt settlements) and communal wastes (garage dumps), and for fertilizers, agricultural chemicals, and chemical agents

Industrial enterprises such as individual plants, sewage-disposal systems, and injection wells also may be sources of groundwater pollution.

Agricultural Development

Agricultural activities bring about considerable variations in groundwater quality and quantity; the main factors of this effect are:

- Groundwater recharge resulting from seepage losses from the main and secondary irrigation canals and infiltration of irrigation water. The artificial groundwater resources formed (additional recharge) play an important role in the total groundwater balance. The additional groundwater recharge in irrigated areas may exceed the resultant groundwater recharge and cause groundwater-level rise and water logging of the irrigated and adjacent areas. Additional recharge in high-rate groundwater abstraction regions may be of considerable importance as a safe-yield source; in some cases, seepage losses from irrigation canals may create conditions for formation of fresh-water lenses in mineralized water zones.
- Supply of a large amount of salt to groundwater with irrigation water as a result of leaching salts contained in the rocks of the vadose zone.
- An increase in evapotranspiration losses of groundwater from the rising water table, leading to a decrease in the resultant groundwater recharge, an increase in dissolved solids content, and soil salinization.
- Groundwater abstraction by vertical and horizontal drainage structures.
- Agricultural use of organic and inorganic fertilizers, pesticides, and insecticides, which are sources of pollutants penetrating the aquifers.
- Transport of pollutants to groundwater from sewage-disposal fields and seepage from storage facilities of animal and poultry farms and factories.

Some of the above factors contribute to greater recharge of groundwater in irrigated areas, and some of them promote groundwater pollution, which, unlike pollution by individual industrial enterprises, covers vast areas. The main agricultural pollutants are nitrogen and iron compounds, as well as pesticides.

Since artificial recharge of groundwater in irrigated regions is important for controlling safe yield, reconstruction of an irrigation system may be the cause of changes in formation and depletion of groundwater safe yield. For instance, an increase in the efficiency of an irrigation system from 0.52 to 0.60 leads to a decrease in artificial groundwater resources (reduced by seepage losses) by almost 30% (Mirzaev and Bakusheva, 1979).

Hydraulic Construction

Construction of reservoirs causes a rise in groundwater level, which is hydraulically connected with surface water. On the one hand, this process leads to the generation of anthropogenic groundwater storage in the coastal zone of reservoirs where rocks in the vadose zone become saturated, and, on the other hand, it is responsible for a decrease in a resultant infiltration recharge because of initiation or intensification of evaporation from the water table. In the majority of cases, hydraulic construction brings about favorable changes in groundwater-development conditions. For example, if, under natural conditions, groundwater occurs in flood-plain sediments composed of thin, fine-grained sands with a small coefficient of permeability, then, after constructing a reservoir and flooding of tidal territories, the thick terrace sediments, having a higher coefficient of permeability, are saturated in the coastal zone. This makes it possible to construct infiltration well fields at sites where their construction was ruled out prior to creation of the reservoir. The assessment of induced water resources in the area of the Kuibyshev Reservoir showed that the average linear modules of water resources increased from 177.7 to 301.6 l per (sq.km) after constructing the reservoir (Predicting the Effect…, 1984). In some cases, construction of reservoirs causes a formation of new anthropogenic aquifers and their water may be used for water supply (Kovalevsky, 1994).

Hydraulic construction may sometimes deplete safe yield because of changing the conditions of groundwater formation. In many regions, water supply wells often tap the groundwater of river valleys where the bedrock aquifer under development is separated from the river by another aquifer confined to alluvial deposits. Here, if the bottom sediments are improved by silt deposition, then the low-flow period will result in a depletion of the storage of the alluvial aquifer that will be replenished in flooding periods. The replenishment depends on the flood intensity. Depleted storage is best recharged at the time of flooding the territories.

Construction of reservoirs upstream of a well field may substantially degrade the conditions of safe-yield formation because in this case, the time, number, and intensity of floods will change. For example, in some well fields in the Seversk Donets River Valley, the groundwater level declined constantly at a fixed water abstraction rate after constructing a reservoir that prevented floods from reaching the stretch where these water intakes were located. This resulted in a decrease in well-field production and a deterioration of groundwater quality (an increase in dissolved solids content and total hardness) due to the inflow of brackish water from the valley slopes. Safe groundwater yield may also be depleted as a result of flooding the productive aquifer in alluvial deposits by the reservoir water.

Construction and operation of nuclear power plants also have a considerable effect on hydrogeological conditions including total groundwater balance changes, and heat and radioactive pollution of groundwater.

From the above brief description of the types of human impact on groundwater quantity and quality, it follows that, in some cases, groundwater recharge, artificial storage, and safe yield increase. In other cases, water-bearing rocks are drained and safe yield is depleted. For this reason, it is necessary to take into consideration human impact on groundwater resources, primarily in solving problems of groundwater use in the future. It should also be taken into account that changes in hydrogeological conditions under human impact bring about changes in other environmental components (surface water, general landscape conditions, etc.).

An increase in groundwater recharge by surface water leads to a decrease in surface runoff. A decrease in groundwater discharge by evaporation from the water table and by transpiration may result in landscape oppression. In this respect, a safe yield should be estimated when the effect of both engineering factors on formation of groundwater resources and groundwater withdrawal on the other environment components is taken into account. Improving the methods of safe yield assessment for maximum utility of groundwater recharge and discharge under development conditions is an important problem for future investigations.

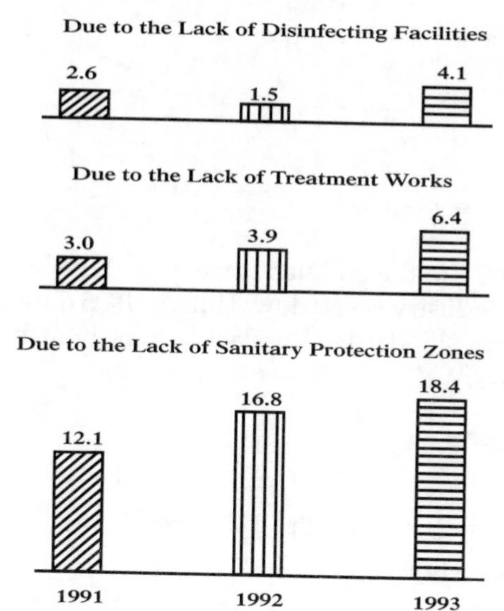

FIGURE 5.1
Centralized sater supply not corresponding to sanitary standards in Russia (in percent).

5.3 Groundwater Pollution in Russia

5.3.1 River Pollution

As mentioned above, groundwater contamination is a serious danger in both highly developed and developing countries. Unfortunately, countless examples can be given, when contamination of water resources (rivers, lakes, reservoirs, groundwater aquifers) makes their use impossible (primarily for drinking purposes), which can cause different negative consequences in the environment and dangerous infectious illnesses.

Some data, characterizing surface and ground water quality, are available for Russia. Intensive human activities in water-catchment areas have brought about irreversible losses and regime changes in river runoff. In the last 10–15 years an increased degradation of small rivers and water-reservoir pollution has been observed. As shown in Figure 5.1, an increase in water-resource contamination during a 3-year period was caused by the absence of sanitation zones and defects in decontamination.

Development has increased the volume of poorly treated wastewater that has caused pollution of natural, primarily surface, water. Pollution in the last quarter of this century makes this water useless not only for domestic and potable-water supply, but also for technical purposes. Treatment by the available methods also appears to be highly expensive.

As an example, data are given for 1995 by the Committee for Water Resources of Russia. In this year, only about 9% of the required volume was decontaminated (purified) according to the standard in the Russian Federation. Most of the purifying facilities (sewage-purifying works) cannot provide adequate sewage purification because of their obsolescence, irrational location, inadequacy of purifying technology for sewage water, excess of contaminant concentration in sewage water, an overloading by the volume of sewage water, etc.

The main sources of water-resource contamination in Russia are as follows:

- housing and public services

- agriculture
- chemical and petrochemical industry
- power engineering
- forestry
- woodworking
- wood-pulp and paper industries

These sources dispose 80% of the polluted wastewater. The level of wastewater purification in all the branches of industry is very low. Thus, in 1995 only 11.6% of industrial wastewater was purified up to the standard in Russia. For some branches of industry, the percent of purification is as follows:

- power engineering–13.7%
- fuel–18.4%
- forestry, woodworking, wood-pulp and paper–2.4%
- materials for construction–10.3%
- electricity–10.9%
- food–3.4%
- ferrous metallurgy–11%
- nonferrous metallurgy–17.3%
- chemical and petrochemical–13%
- machine-building–10.3%

On the whole, water quality of the main rivers of Russia remains unsatisfactory for most purposes, i.e., potable, domestic, public, fishery, cattle-breeding. Management of water resources becomes more complicated with every year. The water-management balance of some river systems (the Volga, Don, Kuban, Terek, etc.) is very strained, and water is actually highly contaminated everywhere. About 30 mln. t of contaminants are disposed into rivers and lakes (according to the State Committee for Statistics data), with sulfates and chlorides accounting for the largest part (about 90%). Most rivers and lakes are heavily contaminated with public and domestic wastes. As a result, surface-water quality does not satisfy sanitary and hygiene requirements.

The State Committee for Sanitary and Epidemic Control, Russian Federation's experts' assessment of sources of surface water supply in different areas of the country is given below.

In the Caspian Sea basin the Volga reservoirs are polluted with phenols (3–5 times the MPC), copper compounds, ammonium and nitrate nitrogen, petroleum products, etc. The Oka River is mostly contaminated in the area between the towns of Serpukhov and Kolomna. In the Kama basin, the Chusovaya River is polluted (compounds of copper and zinc, 6-valent chromium, manganese are 1–5 times the MPC). In recent years, due to development of the Astrakhan Gas-Condensate Complex, surface-water contamination in this area increased, particularly of the Akhtuba, Berketa, and Kigacha rivers, which are sources of potable and domestic water supply.

A very difficult water-management situation is in the Terek River basin. Its water is polluted with copper, zinc, ammonium, and nitrate nitrogen, and petroleum products.

The main sources of surface-water pollution in the Don River basin, where domestic water use reaches 64% of an average annual runoff, are wastewater from industrial projects, many mines, agricultural and cattle-breeding complexes, and also irrigation systems and farms. The Don River is polluted with copper compounds, nitrate nitrogen, petroleum products and phenols, exceeding MPC by tens of times, and in the lower flow (course)–by chlororganic pesticides and metaphos. The Kuban River is polluted with petroleum products, copper compounds, and nitrate nitrogen.

In the Neva basin, the rivers Okhta (in the area of St. Petersburg), Carpovka and Slavyanka are polluted (oxygen deficit, petroleum products and phenols–more than 10 times the MPC). In Ladozhsk Lake, which is a source of potable and domestic water supply, the water in the coastal region of its northern part is most heavily polluted (petroleum products and phenols–more than 10 times the MPC).

Small rivers of the Kolsk peninsula are highly contaminated. Average annual concentration of copper compounds and nickel in the Nyudual and Kolos-Iioki rivers exceeds MPC by tens and hundreds of times. In the North Dvina River basin, its tributary the Puksa River is highly polluted (up to 50–100 times the MPC) by ammonium nitrogen, phenols, lingosulfonates.

Even in Siberia and the far eastern part of Russia, where engineering and economic activities are not so intensive and widespread, surface water is considerably polluted. Thus, the Ob' River, from its source to its estuary, is contaminated with petroleum products and phenols, concentrations being from 5–17 times the MPC. In the Tol' River in the area of the town of Kemerovo, specific highly toxic matters are recorded (aniline, caprolactam, formaldehyde, and methanol).

Average annual concentrations of petroleum products, ammonium nitrogen, copper compounds, and zinc in the Iset' River downstream from Ekaterinburg exceed the MPC by tens of times.

The Enisei River, below Krasnoyarsk, is contaminated with lignosulfonates and volatile acids. Bratsk and Ust'-Ilimsk water reservoirs are highly polluted with wastewater of wood-processing complexes (concentrations of metyl-mercaptan and hydrogen sulfide amount to hundreds of times the MPC).

In the Amur River in the area of towns Blagoveshchensk, Khabarovsk, Amursk, and Komsomol'sk-na-Amur, the content of many pollutants is 1–4 times the MPC, and copper compounds and 6-valent chromium is up to 5–15 times the MPC. In Primor'e, the Rudnaya River is poorly contaminated by boron-containing matters and metal compounds. Concentrations of copper and zinc amount to 30–60 and 80 times the MPC, respectively. Rivers of the northern part of Sakhalin (Okhtinka, Ekhabi, Erri, Pil'turn, etc.) are polluted with wastewater from the oil and gas industry.

When speaking about potable groundwater quality, it should be noted that about a half of the population in Russia must drink water that does not satisfy hygienic norms. Thus, every eighth water sample by bacteriological indicators and every fifth by chemical ones do not correspond to the standard. As shown in Figure 5.2, most areas of Russia show water samples with unfavorably high mircobiological content. About a half the population of Moscow use substandard water from time to time. Near Moscow, about 12 towns have consistently shown territorial pollution that is ten times the background values.

The above-mentioned facts allow us to make a pessimistic conclusion about surface water quality and contamination in many regions of Russia. For this reason, fresh groundwater aquifers have still more value, since they are primarily protected from hazardous contamination.

5.3.2 Quality and Pollution of Groundwater

The general characteristics of groundwater quality and its tendencies for change under the impact of both natural and mainly anthropogenic factors are given below.

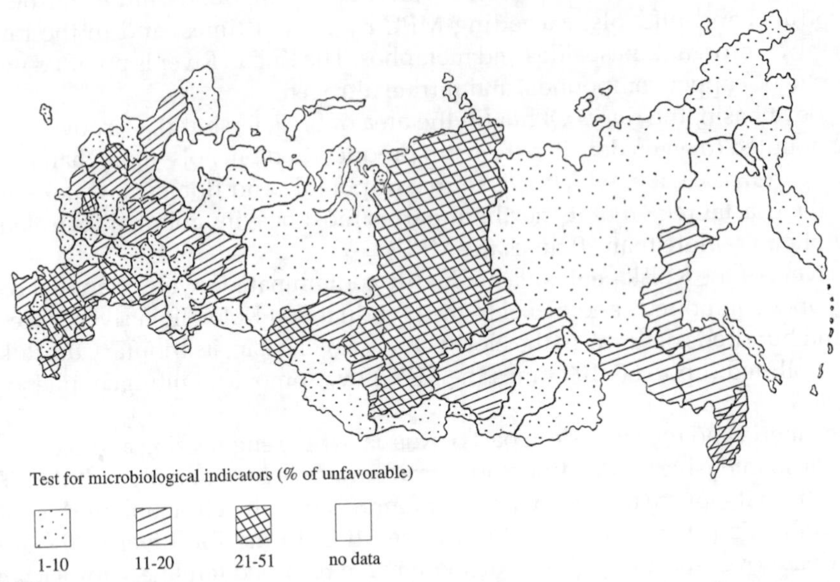

FIGURE 5.2
What water does Russia drink? (according to the State Committee for Sanitary and Epidemic Control data).

On the whole, groundwater quality in the territory of Russia changes within a wide range, following certain regularities. The main regularities for the formation and distribution of groundwater with a certain chemical composition and quality are caused by horizontal and vertical hydrochemical and hydrodynamic regions.

Horizontal issues are mainly characteristic for the upper aquifers and primarily for groundwater. In the plains, it is manifested by increased general mineralization from the north to the south. In northern regions of the country fresh, mainly hydrocarbonate water occurs to depths of about 300 m, and in the south, it is replaced in wide areas by brackish and saline water of chloride composition, with local areas of fresh groundwater.

The vertical zonality of groundwater is observed in artesian basins, where the general mineralization of groundwater and its chemical composition changes with decreasing water-exchange intensity. In the most common case, three hydrogeochemical zones are singled out from the top to the bottom in a vertical cross section of the artesian basin: a zone of fresh (up to 1 g/l) mainly hydrocarbonate water, a zone of brackish and saline (1–35 g/l) sulfate and chloride water, and a zone of chloride brines (more than 35 g/l). This general scheme can be considerably changed in different basins: In some cases, only two out of the above-mentioned zones occur in the cross section, and sometime, only a freshwater zone is found. In some basins of the arid zone, a reverse hydrochemical zonality is often observed. Here, freshwater horizons occur below brackish and saline groundwater or alternate with it. It should be noted that this general scheme of hydrogeochemical regularities is disturbed by the influence of different local factors: lithological (for instance, occurrence of salt-bearing or gypsum-bearing rocks in the upper zone); tectonic (causing an inflow of highly mineralized water from below, along the faults, into a fresh water zone or vice versa of the lower horizons in tectonic disturbances zones); hydrodynamic (causing leakage and close hydraulic connection of aquifers); and also anthropogenic factors.

The role of anthropogenic factors in the formation of groundwater composition and quality, and primarily fresh groundwater used for water supply, has significantly increased in recent decades. At present, the notion of ecologically pure groundwater requires considerable correction in some regions, because of increasing environmental groundwater pollution. In Russia, great attention is paid to preventing not only groundwater-resources

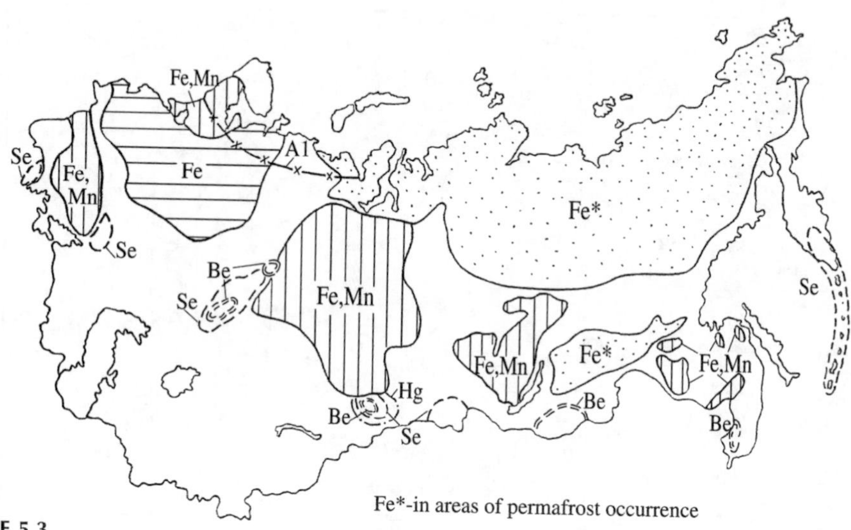

FIGURE 5.3
Hydrochemical provinces of groundwater with raised Fe, Mn, Al, Se, Hg, Be in the territory of Russia and surrounding countries.

depletion (observation of well-field yield, water level in exploitation of adjacent aquifer, etc.), but also to control groundwater quality in exploited horizons.

Groundwater pollution, occurring in recent decades in many regions, is a serious hazard, essentially limiting the possibilities and perspectives of groundwater use for a potable water supply. An increase in concentration of compounds of nitrogen, iron, manganese, strontium, selenium, arsenic, fluorine, beryllium and organic matter is most often observed in groundwater and makes it useless for drinking purposes without special treatment.

The formation of substandard, and in some cases economically hazardous, groundwater occurs not only as a result of its contamination with industrial, agricultural, public and domestic wastes, but also under the impact of natural geochemical processes. These processes cause formation of vast hydrogeochemical provinces, where groundwater is enriched by one or several normalized components. In Figure 5.3, there is a schematic map of groundwater hydrogeological provinces in Russia, published in the work entitled *Planning and Managing the Human Environment*. There are regions on the map of Russia that show an increased content of some components. This work also shows that groundwater quality is most unfavorably affected by contamination from both the geological environment and groundwater itself. The reason that groundwater contamination is a threat to potable and domestic water supplies is, in many cases, petroleum-products leakage from gasoline tanks and pipelines. For example, in the territory of Russia, more that 100 sources of groundwater contamination have been revealed (mainly sulfates, chlorides, nitrogen compounds (nitrates, ammonia, ammonium), petroleum, phenols, iron compounds, heavy metals (copper, zinc, lead, cadmium, and mercury). Areas of groundwater pollution amount in some cases to tens and even hundreds of square kilometers. Most large hearthes are revealed in Moscow, Tula, Volgograd, Kemerovo, Perm, Chelyabinsk regions, in Tatarstan and Bashkortostan (Kochetkov, Yazvin, 1992). Groundwater pollution in these operating well fields is most hazardous. At present, groundwater pollution is registrated at about 140 well fields that supply 87 Russian towns with water.

Figure 5.4a–Figure 5.4c show the main groundwater pollution areas in the European part of Russia, where existing norms of some components in the groundwater are considerably exceeded. The excesses are observed for both natural components of chemical composition

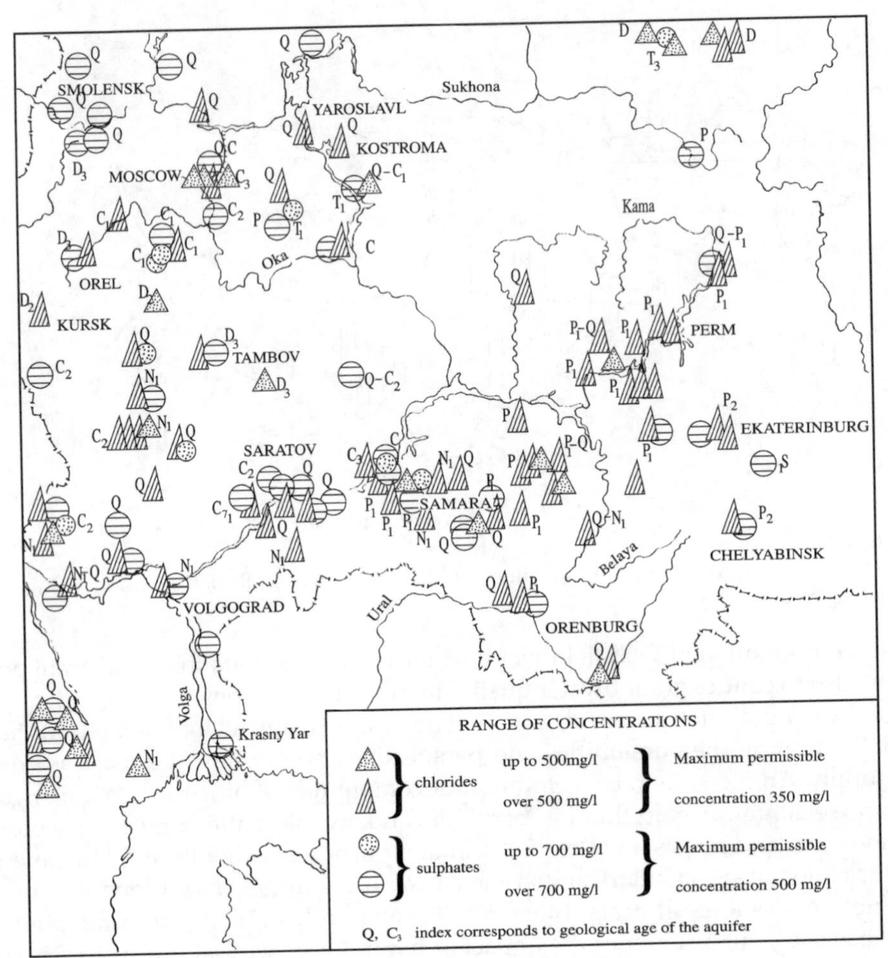

FIGURE 5.4A
Groundwater pollution of the Central European part of Russia: range of concentrations.

and component production, which is caused by human additions such as oil products, pesticides, etc.

Industrial plants are the main source of groundwater contamination, amounting to 42% of all the contaminated sites. This is followed by waste accumulators and filtration fields, wastewater irrigation from cattle-breeding farms, filtration from agricultural fields where pesticides, manures, and fertilizers are used (20%). Fourteen percent of the sites are contaminated with wastewater and public-service wastes. Non-standard groundwater also serves as a source of contamination because of its leakage to the well fields when the production is disturbed (Figure 5.5).

Well fields in Cherepovets (phenols, chlorbenzene, toluol), Lipetsk, Tula, Voroniezh, Tolyatti, Volgograd, Stavropol, Mozdok and Grozny (oil products), Chelyabinsk (phenols, lead, iron), Novokuznetsk (phenols, fluorine), Abakan, Angarsk (oil products) etc., are most heavily contaminated.

Some towns in Russia (Nizhniy Tagil, Kamenetsk-Uralsky, Orsk, Bratsk, Cherepovets etc.) where water resources, including groundwater, are heavily contaminated are declared to be zones of ecological disaster. There, concentrations of heavy metals, oil products, fluorine, and other components are ten times higher than the maximum permissible concentration (MPC).

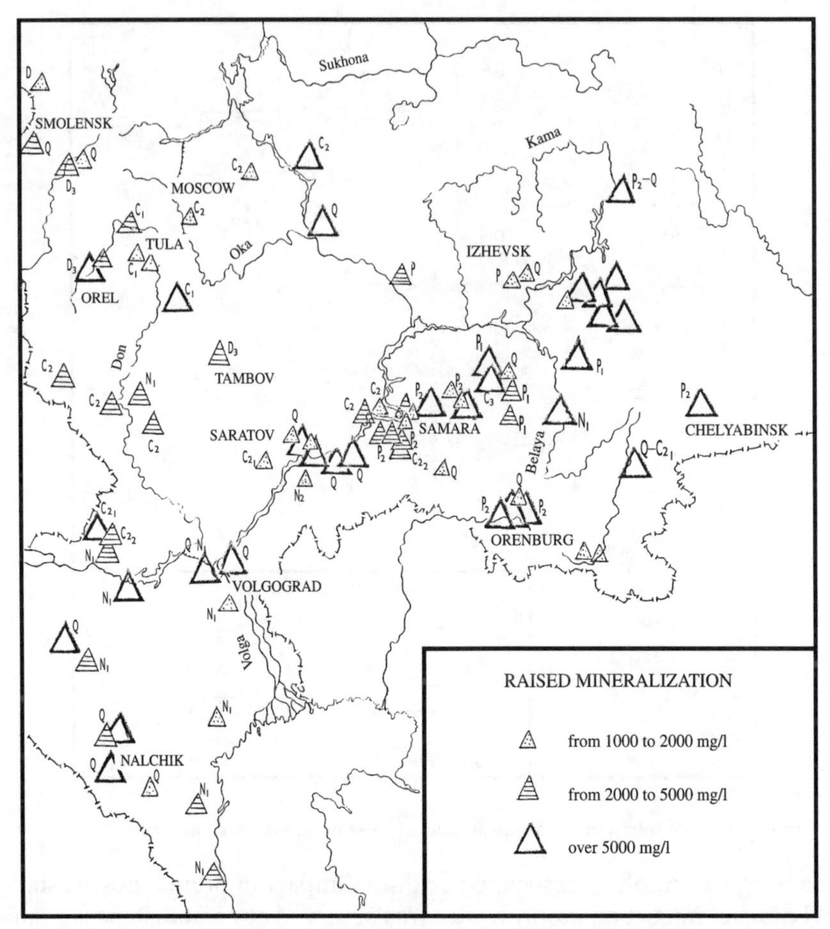

FIGURE 5.4B
Groundwater pollution of the Central European part of Russia: raised mineralization.

In addition to groundwater pollution in separate wells and well fields, regional groundwater pollution occurs. Regional changes of groundwater composition and properties are usually caused by both point and areal pollution sources.

Close interaction of groundwater with the environment and other components has been especially well demonstrated, as was obvious in recent decades, when the impact of man-induced factors on the environment progressed greatly.

For example, urban impact on groundwater quality on a regional scale is the most intensive. This results from increased mineralization of precipitation in urban areas and "acid" rain, oil-products leakage, and the impact of industrial and sewage waste. Groundwater salinity in urban territories is usually 2–3 times higher than in rural areas.

The impact of acid rain is the second example. It is known that atmospheric emission of chemicals doubles every 10 years, which results in an increase in their concentration in atmospheric precipitation. Under infiltration of atmospheric precipitation and snow-melt water, many different elements that change the groundwater's hydrochemical regime, its composition and quality enter. The content of heavy metals in the groundwater essentially increased in the areas of infiltrating acid atmospheric precipitation. For instance, concentration of aluminum, zinc, and manganese in the snow cover in Moscow increased 15–20 times over the last few years.

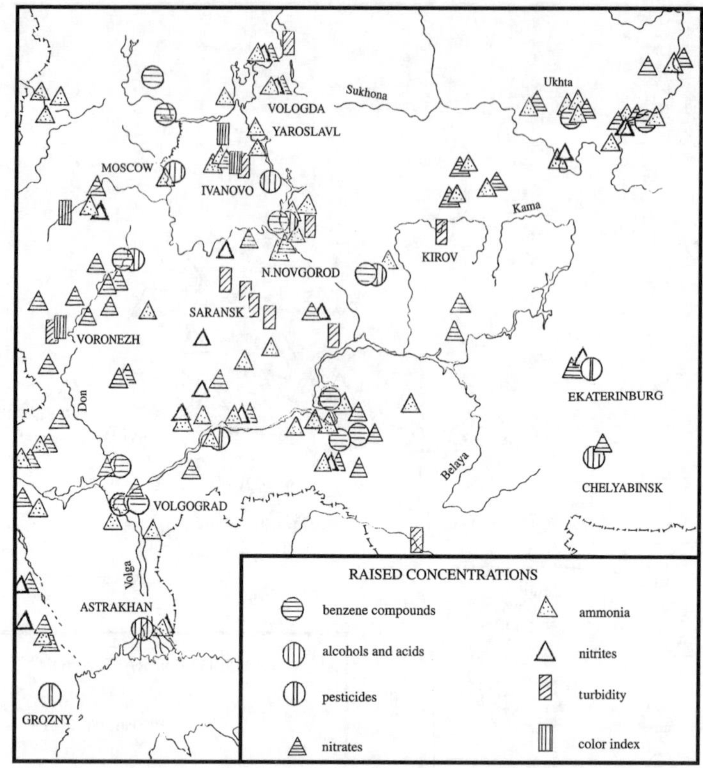

FIGURE 5.4C
Groundwater pollution of the Central European part of Russia: raised concentrations.

The most tragic example is associated with the impact of atomic power stations. The Chernobyl disaster impact on groundwater was observed at great distances — up to 100 k from the site of the catastrophe. Considerable radionuclide accumulations were formed in the upper soil level in the Chernobyl area. Radionuclide migration through the vadose zone resulted in the growth of their concentration several tens and even hundreds of times and even at large depth (up to 100 m), if compared with the situation before the failure.

There are many similar cases. All these examples show that regional environmental pollution results in regional groundwater pollution. This makes it clear that problems of groundwater protection from contamination are closely related to a general problem of environmental protection from contamination.

Hazardous environmental and groundwater contamination in the Volga basin was described earlier. Some more pronounced examples will be given here, though they are far from a full characterization of groundwater pollution in Russia and the measures taken to protect groundwater.

Oil products removed from the soil and groundwater is an important way of eliminating oil contamination from oil tanks and pipelines. A lens of oil products exists in Grozny (Chechnya). Its thickness is 3 m and it spreads over 1.5 km^2. To study conditions of lens spreading and ways of its elimination, the Russian firm HYDEC (Hydrogeological Research and Design Company) made a series of drilling and test filtration studies and developed a model of water- and oil-product filtration under local hydrogeological conditions. The volume of oil product of the delineated part of the lens amounts to 700 000 t. Special variants were suggested to intensify oil products removed to localize the lens and eliminate it.

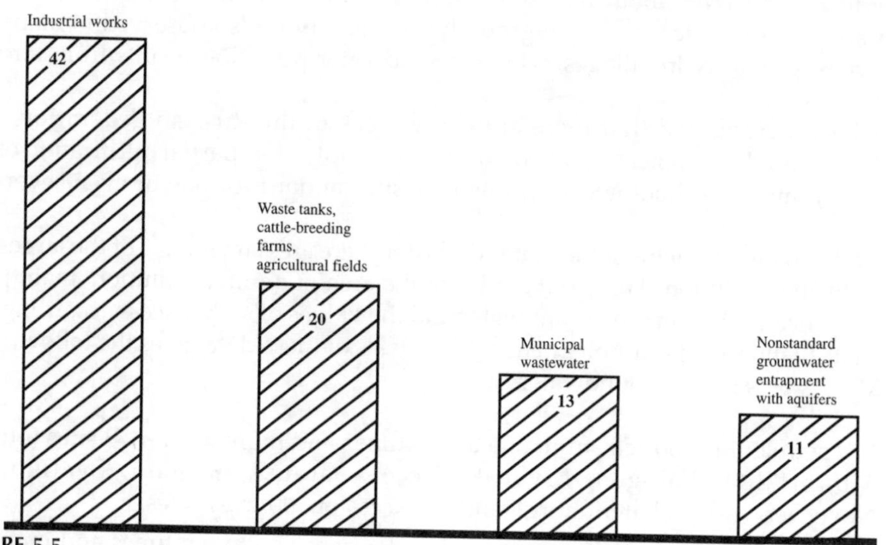

FIGURE 5.5
Sources of groundwater contamination in Russia (percentage of the total contaminated sites).

Much smaller in size but not less dangerous, a lens of kerosene was formed as a result of oil-products leakage in Eysk, on the shore of the Taganrog Bay in the Azov Sea. The lens amounts to 0.7 km^2 at a depth of 20–22 m. The hazard is, which is driven by hydrogeological conditions, has caused polluted groundwater and the lens to move toward the Bay. Since 1990, kerosene seepage that occurred along 700–800 m of the shoreline on Taganrog Bay caused the seawater pollution. The square of kerosene spot on the sea surface is about 400 m^2. Measures to protect the coast are utilized here and will help to prevent kerosene intrusion into the sea. Work to eliminate contamination is under way, and is expected to take about 10 years.

In some cases, groundwater contamination is a threat to river-water quality. The town of Uhta (Komi Republic) can serve as an example. Oil-products leakage from oil refineries and storage areas has caused surface soil pollution as well. Oil products infiltrate through the vadose zone and accumulate in the groundwater surface in the aquifer. The area of contaminated groundwater covers about 5 km^2. Groundwater containing oil products is drained by the Uhta River and also contaminates it. To make recommendations on preventing the river contamination, the sources of contamination were investigated. Also, special wells were drilled, test filtrations were carried out, chemical analyses of water and soil were made, and a numerical model of groundwater filtration in the river basin was developed by the VODGEO Institute. As a result, the source of contamination, its area, and intensity were determined and measures were worked out to save the river. These measures include linear horizontal drainage combined with antifiltration wal, and removing oil products from the groundwater surface.

Nitrate contamination, caused by fertilizer applications, occupies an appreciable place in the problem of soil and groundwater pollution. Thus, groundwater contains up to 100 mg/l of nitrates, which is 10 times the MPC in the Moscow region.

Nitrogen introduction to soil and groundwater is considerably less in Russia than in other European countries because manure applications have been substantially reduced from 1990–1994 and now do not exceed 50 kg/ha a year. Thus, nitrogen contamination, caused mainly by cattle farming and domestic waste, is not widespread.

Based on the data of the State Water Cadaster (SWC) for 63 towns and settlements of Russia, a preliminary analysis has been made of groundwater pollution in the main aquifers.

According to the SWC, about 1000 sites of groundwater pollution have been considered in Russian territory. Deterioration of groundwater quality is also observed in many individual wells, primarily in villages, where groundwater pollution results from farm and domestic wastes.

It has been determined that there are many areas in the Russian Federation where groundwater, used for domestic and potable water supply after partial purification or even without preliminary technological treatment, is substandard for one or even several elements.

In the Information Bulletin of the Water Cadaster there are data for 516 of the largest sites of groundwater pollution. These data indicate the area of aquifer pollution, main pollutants, their concentration in the groundwater and the values of MPC excess according to All Russia State Standards (pollution intensity). Analysis of these data gives the following values of MPC excess in the groundwater:

- For petroleum products, from one to 100 times and more in over 73 sites (out of 516) (Ukhta, Vologda, Bryansk, Oryol, Saransk, Samara, Pyatigorsk, Novocherkassk, Yekaterinburg), and in excess of 100 in 17 sites.
- For nitrates and other nitrogen compounds, from one to ten times and more in 102 sites (Verkhne-Dneprovsk district of Smolensk region, Nereksk district of Kaluga region, Kulebaksk district of Nizhegorodsk region, Novokuznetsk, Novosibirsk, Yuzhno-Sakhalinsk, etc.), and in excess of ten times in 32 sites of groundwater pollution.
- For phenols, from one to 100 times and more in 41 sites (Yemva in Komi, Sovetsk in Kaliningrad region, Cherepovets in Vologda region, Redkino in Tver region, Moscow, Lyubertsy and Staraya Kupavna in Moscow region, Dankov in Lipetsk region, Volgograd and Volzhsky in Volgograd region, Saratov, Tolyatti, Grozny, Ufa, Tomsk, Kemerovo, Angarsk, etc.), and in excess of 100 times in 17 sites.
- For iron compounds, from one to 100 times and more in 65 sites (Yardage in Komi, Plesetsk district in Archangels region, Pskov, Ivanovo, Tambov, Aktyubinsk, Chelyabinsk, Chita and Korymsk district in Chita region, Khabarovsk, etc.), and in excess of 100 times in 12 sites.
- For sulfates, in 13 sites of groundwater pollution, including up to five times in ten sites and in excess of five times in three sites (Kineshma in Ivanovo region, Dzerzhinsk in Nizhegorodsk region, Volgograd, Balakovo in Saratov region, Krasnoyarsk, etc.).

The general mineralization of groundwater in areas of pollution exceeds the MPC in 25 sites (out of 516), including several times the MPC in 17 sites (Voskresensk in Moscow region, Oryol, Skopin Ryazansk region, Smolensk, Tula, Ukhta, Tambov, Saratov etc.), and in some tens of times in eight sites (Zavolzhsk in Ivanovo region, Shchebekino in Belgorod region, Bereznyaki in Perm region, Dzerzhinsk, Neftekumsk in Stavropol region, Nalchik, etc.).

There are 30 sites where groundwater hardness exceeds MPC by several times (Oryol, Smolensk, Tula, Krasnoyarsk district in Astrakhan region, Volzhsk district in Volgograd region, Samara, Neftekumsk, etc.), and in some tens of times in five sites (Volgograd, Solikamsk and Bereznyaki in Perm region, Chelyabinsk, Tuimazinsk district in Bashkiriya).

In seven sites, the content of different organic compounds in the groundwater exceeds MPC by more than 100 times and it is excess of 100 times in 41 pollution sites.

Groundwater bacterial pollution is often observed, particularly from the first water removal from surface aquifers.

It should be noted that pollution of aquifers, which are intensively used for water supply for cities and towns, is the most dangerous, particularly in places where groundwater is often the main — even the only — source of domestic and potable water supply. Therefore, the analysis of groundwater quality used for these purposes was made in 63 towns and settlements in Russia. The exceedance of the All-Russia State Standards was observed: for iron–in 47 towns, for hydrogen sulfide–in three, for hardness–in 18, for phenols–in 16, for barium–in four, for chromium–in four, for lead–in two, for turbidity–in 18, for arsenic–in three towns, etc.

Table 5.1 shows towns where the content of certain elements in exploited aquifers exceeds MPC by several times.

In recent years, the problem of groundwater protection received great attention in Russia, as in many other countries. Different means of protection are used, including development of monitoring the vadose zone and aquifers, designing regional permanent analogies,

TABLE 5.1

MPC Excess in Aquifiers of Selected Towns

Element	Number of towns and settlements	All-Russia State Standard for MPC	Towns, MPC excess
Iron	47	0.3	Ivanovo–in 2 times, Lipetsk–1.5, Kropotkin–3, Pechora–4, Kursk–2, Balakovo–3, Orenburg–4, Yarozlavl–2, Pskov–2, Berezniki–2
Manganese	15	0.1	Kirillov–in 1.2 times, Baltiisk–2, Kursk–2, Ivanovo–2, Orenburg–2
Fluorine	5	1.2	Kanashi, Chuvashiya–1.5 Kovylkino, Mordoviya–1.2, Kamenka, Penza reg.–1.3, Kulebaki, Gorky reg.–1.3, Archangels–1.3
Water is subjected to fluoridation	24		
For hydrogen sulfide	3		settl. Komsomolsky, Kalmykia; settl. Alekseevskt, Volgograd reg.; t. Borisoglebsk
For mineralization	5	1000 g/l	t. Balakovo, Saratov reg.; settl. Semikarakor, Rostov reg.; settl. Komsomolsky, Kalmykia; t. Orenburg
For turbidity	8	up to 1.5 mg/dm3	t. Ivanovo–in 2 times; Ostrogozhsk–1.5; Kirov–1.5; Orenburg–1.5; Archangels–1.5
For color index	4	20 degree	t. Ivanovo–1.5 times; Rostov, Yaroslavl reg.–3; Oryol–1.2; Ostrogozhsk–1.2
For hardness	18	up to 7 mg-equiv./l	t. Kursk–1.5 times; Rostov, Yaroslavl reg.–3; Oryol–1.2; Ostrogozhsk–1.2; Kursk–1.1; Borisoglebsk–1.5; Novotroitsk–1.4; Buguruslan–1.2; Orenburg–1.2; Krasnodar–1.1; Voronezh–1.5; Novgorod–1.6; Sterlitamak–2
For chlorides	7	up to 350 mg/l	settl. Alekseevskii, Volgograd reg.–2 times; Voronezh–1.1; Semikarakor, Rostov reg.–1.5; Pskov–1.2

Using such water for drinking purposes can cause different illnesses. This aspect is considered in detail in Chapter 8.

providing zones of aquifer sanitary protection, experimental studies of contaminant filtration and migration, etc.

Groundwater quality remediation in Russia consists mainly of protecting well fields and surface water streams and reservoirs from contamination. The technique of calculation and prediction of mass transfer in polluted aquifers including the use of analytical methods and methods of modeling are well enough developed.

6

Groundwater Vulnerability

6.1 Regional Studies State of the Art

Groundwater in many regions is the most important, ecologically safe and often the only source of potable water supply.

However, substantial human impact on the environment, including groundwater as an environmental component, has resulted in a situation where groundwater cannot be used for water supply in some regions. This particularly concerns shallow groundwater of the upper hydrodynamic zone.

Groundwater quality change under human impact is often expressed as an increase in dissolved-solids content and contents of some chemical elements and compounds (chlorine, sulfates, calcium, magnesium, iron, fluorine, and some others), in the appearance of toxic substances of artificial origin (e.g., pesticides, oil products, and radionuclides), in a change in temperature and pH, and in the appearance of odor and color, etc. In many areas, the human impact on groundwater has acquired a regional character that is due to general pollution of the environment, particularly pollution of the atmosphere, soil, and surface runoff. The degradation of groundwater quality is particularly acute in industrial regions and in areas with high-rate application of chemical fertilizers. Groundwater-quality deterioration from pollution frequently constitutes a graver danger than actual water shortage. Under these conditions, groundwater protection from pollution becomes one of the principal tasks of modern hydrogeological science and practice.

The concept of groundwater vulnerability is based on the assumption that the geological environment can provide a certain degree of groundwater protection from natural and anthropogenic impact, particularly from contaminants available in the soil-rock zone. The term "groundwater vulnerability" was introduced by French hydrogeologist J. Margat (1968) at the end of the 1960s. The first synoptic map of an aquifer vulnerability to pollution at a scale of 1:1 000 000 was published at the beginning of 1970 (Albinet, 1970).

Although the concept of groundwater vulnerability has been existing for the past three decades, there is still no exact definition for it. However, the following is often used: "By groundwater vulnerability we mean those groundwater system natural properties that depend on the sensitivity of this system and its ability to cope with natural and anthropogeni impact" (Vrba and Zaporozhets, 1994).

The Committee of Technology for estimating groundwater vulnerability under the National Investigation Committee of the USA (1993) defined groundwater vulnerability as a tendency or probability of achieving by a contaminant a certain position in a groundwater system after getting into/above the aquifer zone. However, the Committee later singled out two main types of vulnerability: specific ones (certain contaminant or a class of contaminants), and inherent ones, which depend on the properties and the behavior of specific contaminants.

The basis of the groundwater vulnerability concept is in understanding the fact that in some areas, due to peculiarities of the natural conditions (primarily, geological-hydrogeological ones) groundwater is easily affected by contaminants and thus is more vulnerable. Therefore, the main aspect for assessing the natural protection and vulnerability of groundwater is the characteristics and analysis of natural peculiarities in the region under study. Based on them, it is easy to formulate the following definition. The extent of groundwater protection from pollution is the property of a natural system that it makes possible to preserve for a predicted period, and the composition and quality of groundwater in keeping with the requirements for its practical use.

The opposite notion is groundwater vulnerability. The larger the extent of groundwater protection, the smaller groundwater's vulnerability to pollution.

Assessment of natural groundwater vulnerability to contamination is a hydrogeological assessment of measures for groundwater protection under different natural and man-induced conditions. According to the experience of some countries, such as Russia, USA, Germany, and Italy, it is possible to make regional assessments and map the natural vulnerability of aquifers used for water supply and irrigation. This assessment is usually based on the analyses and processing of all the available hydrogeological data and the data that characterizes the protective properties of the vadose zone.

Assessing groundwater's vulnerability to contamination is made in two ways:

1. A quantitative assessment of the territory is made according to the intensity of the impact of different natural and man-induced factors on the aquifers' vulnerability. This facilitates comparison of different parts of the territory from the point of view of their vulnerability.
2. A quantitative assessment of the time (rate) for a certain possible contaminant to penetrate into the aquifer, accounting for natural properties of the water-bearing and overlaying rocks and the migration abilities of the contaminant.

In other words, there are two different approaches: the first is assessment and mapping of groundwater protective properties or vulnerability of any territory without taking into account characteristics and properties of certain contaminants. The second is assessing and mapping protective properties of a natural system as applied to a certain type of contamination.

Let us consider, in short, the present methodical approaches to assessing and mapping groundwater natural protection. As mentioned, studies in this direction in the last few years in a number of countries have been more frequent.

The first effort in the development of methodological approaches to the evaluation of groundwater vulnerability were made in the late 1960s, and at the present time, there is a great variety of different approaches to the assessment of groundwater vulnerability. The relevant maps are based on approaches differing in scale and content (Table 6.1). A detailed analysis of all existing methodologies is not attempted in this book. However, the main trends in their development have been considered.

Most of the methodologies are based on either quantitative or qualitative (using computation formulas) conditions that consider the effect of some factors on groundwater vulnerability. An intermediate position involves the determination of the degree of protection by a sum of numbers corresponding to the contribution of a certain factor. The methodologies based on this principle can be grouped with semiquantitative approaches. The quantity and types of vulnerability factors considered by different investigations vary. The degree of the protection of unconfined groundwater, or the water of an upper confined aquifer is commonly assessed.

TABLE 6.1

List of Some Published Groundwater Vulnerability Maps[a]

No	Year	Country	Authors	Title	Scale	Reference
1.	1967	Czechoslovakia	M. Olmer	Map of groundwater vulnerability to pollution	1:200000	Sovremennoe sostoyanie (1977)
2.	1967	Poland	Anon.	Hydrogeological map of the Alsztin Province with characteristics of groundwater vulnerability to pollution	1:100000	Sovremennoe sostoyanie (1977)
3.	1967	Poland	Anon.	Hydrogeological map with elements of danger of groundwater pollution from surface	1:200000	Vrana (1984)
4.	1968	USA	W.H. Walker	Map of potential pollution of an aquifer in Illinois	n.d.	Vrana (1984)
5.	1968	Czechoslovakia	M. Vrana	Groundwater vulnerability map of Bohemia and Moravia	1:500000	Vrana (1984)
6.	1970	Czechoslovakia	M. Banski	Groundwater vulnerability MPA of Slovakia	1:500000	Vrana (1984)
7.	1970	France	M. Albinet	Map of aquifer potential contamination	1:1000000	Vrana (1984)
8.	1970	Germany, Fed. Rep.	Anon.	Map of the northern Rhine-Westphalia Province with indication of five areas of water purification by infiltration	n.d.	Sovremennoe sostoyanie (1977)
9.	1971	Poland	Anon.	Map of groundwater pollution danger	n.d.	Sovremennoe sostoyanie (1977)
10.	1971	Spain	Anon.	Map of regional subdivision of area of Spain with six main categories of groundwater contamination danger	1:200000	Sovremennoe sostoyanie (1977)
11.	1973	Poland	A.S. Kleczkowski, A. Miszka, Z. Pirit	Map of danger of pollution and protection of groundwater of Poland	1:10000000	Vrana (1984)
12.	1973	France	B. Lemer P. Martin	Map of groundwater contamination vulnerability with indication of contamination sources and surface water contamination degree, area of Monelier, two sheets	1:100 000	Vrana (1984)
13.	1974	France	J. Lavis J. Poutallat	As above, five sheets	1:50000	Vrana (1984)
14.	1975[b]	France	R. Tossin, J. Lienhart, J. Collin	As above, area of Lyons	1:20000	Vrana (1984)
15.	1975	Czechslovakia	M. Olmer, B. Rezak	Map of groundwater vulnerability of Czechia, 18 sheets	1:200000	Vrana (1984)
16.	1976	Czechslovakia	M. Vrana	Legend for large-scale map of groundwater vulnerability	1:20000 1:50000	Vrana (1984)
17.	1976	USSR	N.V. Rogovskaya	Model map of groundwater vulnerability for one of the regions of the USSR	1:200000	Rogovskaya (1976)
18.	1978	Bulgaria	C. Antonov, M. Raikova	Map of natural conditions of degree of protection of groundwater	1:100000	Vrana (1984)
19.	1979	Poland	A. Macioszczyk Z. Plochniewski	Map of groundwater vulnerability	n.d.	Vrana (1984)

TABLE 6.1 CONTINUED

No	Year	Country	Authors	Title	Scale	Reference
20.	n.d.	Germany, Dem. Rep.	Anon.	Map of groundwater vulnerability with indication of five types of regions by self-purification and migration of polluted infiltrating water	n.d.	Pitieva (1984)
21.	1982	Denmark	A. Villumsen O.S. Jacobesen C. Sonderskov	Groundwater vulnerability map for one region of Jutland (methodology involves geological, hydrogeological, and hydrochemical data)	n.d.	Villumsen and others (1983)
22.	1980	USSR	V.M. Goldberg and others	Maps of degree of groundwater protection from pollution for (1) USSR area, (2) European USSR area, (3) Certain Russia Regions, and (4) Moscow region district	1:2500000 1:150000 1:500 000 1:50000	Goldberg (1993)
23.	1983	Germany, Fed. Rep.	Anon.	Map of natural potential of the environment in Lower Saxony and Bremen	1:200000	Josopait (1983)
24.	1988	Poland	J. Gorski and others	(1) Groundwater vulnerability map of Poland (2) Ground water vulnerability maps for each province	1:500000 1:100000	Gorski and others (1988)
25.	1988	USSR	V.M. Matusevich and others	Schematic map of groundwater vulnerability for hydrosphere of permafrost area of western Sibiria	n.d.	Matusevich and others (1988)
26.	1989	Italy	A. Aureli and others	Maps of vulnerability of Faldeidriche, settore Nord Orientale Ibleo (Southeast Sicily)	1:50000	Aureli and others (1989)
27.	1989	Sweden	J. Pousette and others	Map of vulnerability of groundwater of crystalline basement in connection with infiltrating water contamination (supplement to Hydrogeological Map of Jonköping District)	n.d.	Pousette and others (1989)
28.	n.d.	USSR	V.M. Matusevich and others	General map of contamination risk of Ukrainian SSR	n.d.	Drich and others (1990)
29.	1990	Belgium	Anon.	Map of groundwater vulnerability of Flanders	1:100000	Jobe and Gossens (1990)
30.	1990	Germany, Fed. Rep.	Anon.	Groundwater vulnerability map of area of Kreiznach Spa	n.d.	Fürst and others (1990)
31.	1991	Guatemala	Anon.	Maps of aquifer vulnerability of Guatemala	n.d.	Munor and Langevin (1991)
32.	1991	USA	I.S. Zektser L.G. Everett S. Cullen	Groundwater vulnerability map of California	1:2000000	Zektser and others (1991)

[a]The table was compiled using data of O.I. Grozdova (1987).
[b]After 1975, maps of potential groundwater contamination have been constructed every year in France.

In quantitative evaluation of the degree of protection of unconfined groundwater, information concerning the thickness and lithology of the vadose zone, and sometimes data on the lithological character of the water-bearing rocks are taken into consideration. The sorption properties of rocks are usually based indirectly on the lithologic character of the rocks.

The authors of a number of methodologies are concerned primarily with the self-purification effect of groundwater on the degree of its protection from contaminants penetrating through the vadose zone. Specifically, Pitieva (1984) determined protection categories, using a special scheme that considers the capacity of rocks to remove physicochemical contaminants from groundwater as a result of adsorption, ion exchange, sedimentation, and decomposition of organic matter by oxygen and microorganisms. Eight protection categories involving the lithological and mineralogical characteristics, permeability, and thickness of rocks in the vadose zone are identified. The groundwater vulnerability map of Flandres, compiled by Jobe and Cossens (1990), is based on data of the thickness and lithology of the vadose zone and the composition of the water-bearing rocks. The map shows 16 sites characterized by a combination of the three above-mentioned protection factors. For instance, the aquifer sites composed of highly permeable, intensely fractured hard rocks and having a thin (up to 5 m) confining layer are considered most vulnerable. The sites with clayey sand as the water-bearing rocks, and clays as the confining rocks are designated as least vulnerable.

An example of the qualitative assessment of the protection of the upper confined aquifer is the approach taken by Goldberg and Gazda (1984) involving the analysis of two indexes — the thickness of the upper confining layer (m_0) and the position of the levels of the confined aquifer under investigation (H_2) and the overlying unconfined aquifer (H_1). Depending on the combination of these two factors, three classifications of confined groundwater are distinguished. They are 1) protected, 2) conventionally protected, and 3) unprotected.

V. Goldberg (1993) is a co-author of a methodology of semiquantitative analysis involving the determination of the degree of unconfined groundwater protection using a sum of numbers. This method considers the thickness of the vadose zone, the semipermeable clay sediments, and their lithology. A table is used showing gradations of the depths to unconfined water, thickness of semipermeable rocks, their generalized lithology (sandy loams, loams, and clays), and the sum of numbers corresponding to the gradations. The sum of numbers indicates the category of the degree of protection. The larger this sum, the better the groundwater is protected from contamination. This approach has certain evident advantages, including an attempt to evaluate the specific contribution of all indexes under consideration in the evaluation of groundwater vulnerability with the use of Zunker's formula. This methodology is, however, not versatile and its successful application is limited to regions where the geologic structure is complex.

Let us consider in more detail the work of American scientists L. Aller, T. Bennett, J. Lehn, and C. Nackett, who in 1987 worked out a standardized system for regional assessment of groundwater pollution called DRASTIC. They investigated the impact on the environment of the following natural factors:

- depth to a groundwater level and its recharge
- structure and composition of soil and aquifer
- topography
- impact of the vadose zone
- hydraulic conductivity of the aquifer

Each factor is characterized by a constant, previously determined weighting contribution (Table 6.2), for instance, depths to the groundwater level are characterized by weight 5, soil

TABLE 6.2
DRASTIC

Assigned Weights for DRASTIC Feature Feature	Weight
Depth to water	5
Net recharge	4
Aquifer media	3
Soil media	2
Topography	1
Impact of the vadose zone media	5

structure by weight 2, etc. Then a rating of each factor is determined for specific geological-hydrogeological conditions. For example, various depth intervals to the groundwater level (weighting factor 5) are characterized by different ratings, nine depths to 5 m — by one rating, depths from 5 to 10 m - by another one, etc.). By multiplying "weight" and "rating," the authors obtain "number," which quantitatively characterizes the estimated factor on groundwater or its protection. The sum of "number," called a "drastic index," gives the combined impact of all the factors mentioned and is then put on the map. A specific example taken from the authors' work is given in Table 6.3.

TABLE 6.3
SETTING: Sandstone, Limestone and Shale-Thin Soil

General Feature	Range	Weight	Rating	Number
Depth to water table	15-30	5	7	35
Recharge	4-7	4	6	24
Aquifer media	thin bedded ss sequences	3	6	18
Soil media	loam	2	5	10
Topography	2-6%	1	9	9
Impact vadose zone	bedded	5	6	30
Hydraulic conductivity	1-100	3	3	9
DRASTIC				129

Using the above approach, the authors quantitatively assessed groundwater protection from contamination for most regions of the USA and compiled maps with the drastic index indicated. They completed a monumental work on collecting, analysis, and generalization of available real data representing the impact of specific factors on groundwater protection. However, the methods of assessing and mapping groundwater protection suggested in DRASTIC have not been widely adopted, possibly because of certain shortcomings:

- DRASTIC uses a constant value for a weighting factor. These are assigned weights, while the contribution of each factor under natural conditions is different. For instance, using the same depth to the groundwater table affects its protection differently under arid and humid conditions. This circumstance makes the results of assessing protection according to DRASTIC conditional.
- DRASTIC does not differentiate between factors affecting confined and unconfined aquifers This difference will be discussed later.
- It's not quite clear why one soil impact is considered twice, i.e., as the factor "soil media," and as a component of the factor "impact of the vadose zone."

In spite of the above restrictions, DRASTIC should be regarded as the first considerable work aimed at quantitative regional assessment of natural groundwater protection from pollution.

6.2 Principles of Assessing and Mapping Groundwater Vulnerability

Groundwater protection depends on many factors that can be divided into three groups — natural, man-induced, and physicochemical. It is governed by natural factors that include:

- depth to groundwater
- presence of semipermeable layers
- composition, thickness, and permeability of rocks covering groundwater
- sorption properties of rocks
- hydrodynamic conditions determining the direction and velocities of seepage
- rates of water exchange between main aquifers

Man-induced factors comprise:

- presence and storage of contaminants on the land surface
- wastewater disposal
- sewage irrigation
- character of disposal and penetration of contaminants into aquifers

Physicochemical factors include:

- sorption and migration properties of contaminants
- the character for the interaction between contaminants, rocks, and groundwater

All these factors should be considered when assessing groundwater vulnerability to contamination. However, the selection of the principal factors for consideration will depend on the scope of the investigation and the scale of the assessment maps.

It should be noted that natural factors determining groundwater protection or, vice versa, vulnerability to pollution, are different for confined and unconfined aquifers.

For unconfined groundwater those factors are as follows:

- thickness of vadose zone (or depth to the unconfined groundwater level)
- lithological composition of this zone
- groundwater recharge (usually a mean annual value for a perennial period is used)
- residence time of groundwater
- transmissivity of unconfined aquifers

It should be noted that the level of the above factors' impact on groundwater protection differs according to the specific geologic-hydrogeologic conditions of the studied territories.

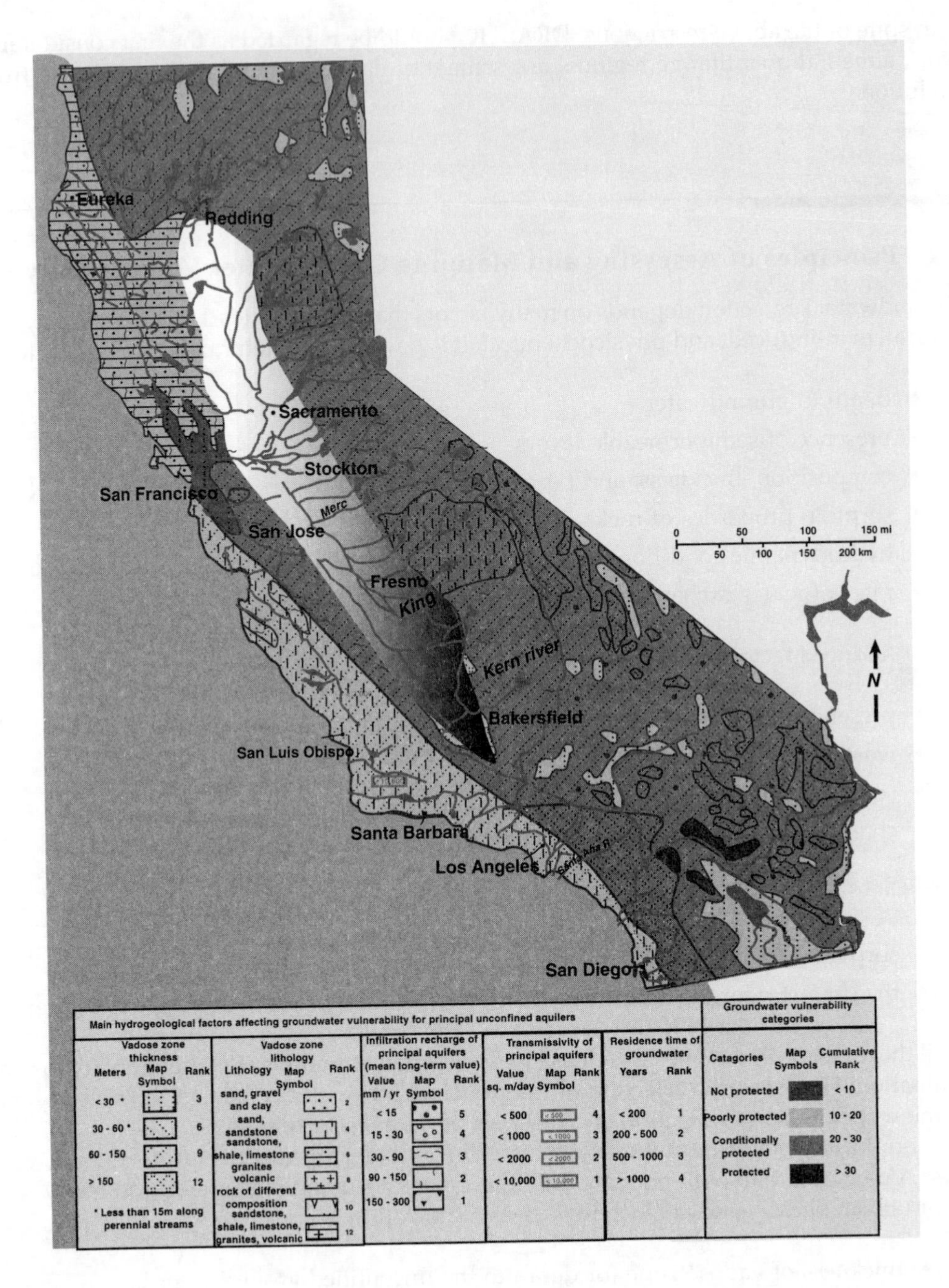

FIGURE 6.1
Groundwater vulnerability map of California. (Prepared by I.S. Zektser, L.G. Everett, S.J. Cullen, T.H. Robinson, and A. Macias.)

Characteristics of vadose-zone structure and permeability are the most important for assessing protection of unconfined aquifers. However, data on permeability of rocks in a vadose zone are often very limited. In this case, for regional assessment, a lithological composition of rocks in a vadose zone is analyzed, and on the basis of this analysis, different rock categories are singled out, e.g., well-permeable, semi-permeable, unpermeable. An important factor that protects natural groundwater from contamination is the volume of its infiltration recharge. For quantitative characteristics of this volume, the results of regional assessment and mapping of groundwater discharge in l/sec per 1 km^2 or mm per year can be used. The map of groundwater discharge in California at a scale of 1:2 000 000 can be used as an example (see Figure 3.2).

Analyzing the impact of natural factors on groundwater protection, I. Zektser, L. Everett, and S. Cullen made a preliminary quantitative assessment of unconfined and partially confined aquifers in California in 1991, and compiled a groundwater-vulnerability map at a scale of 1:2,000,000. The impact of each factor, depending on specific natural conditions, was denoted by different numbers determined in an expert way. For instance, the thickness of the vadose zone (amounting to 30 m) was represented by the number 3; from 30 to 60 m by number 6; the amount of infiltration recharge less than 15 mm per year by the number 10, from 15 to 30 mm per year by 8, from 30 to 60 mm per year by 6, etc. Natural groundwater protection is represented on the map by fixed boundaries (Figure 6.1). Besides the impact of specific factors on groundwater-protection symbols, this map should be considered the first nearly accurate assessment of groundwater protection from contamination in the state of California. On the whole, analysis of different confined and unconfined aquifers makes it possible to conclude that the thicker the vadose zone, the less permeable the lithology, and lower the values of infiltration recharge and transmissibility of an aquifer; the higher the intensity of residence time, the better the natural protection of groundwater and, hence, the lower its vulnerability to contamination.

For confined groundwater. the main natural factors determining its protection from pollution are as follows:

- a ratio of levels (heads) of the assessed aquifer and the overlying unconfined one determining a possibility of polluted water access from above
- thickness and permeability (lithological composition) of the upper confined (waterproof) layer, determining possible volume of polluted groundwater leakage from above
- water transmissibility
- volume of infiltration recharge
- residence time of the assessed aquifer

It is quite clear that the first two factors are decisive, as they together characterize a potential danger of polluted groundwater penetration into a confined aquifer from above. When assessing confined aquifer vulnerability data, a coefficient of leakage should be used. It is known that the leakage coefficient A is a ratio between filtration coefficient (K_0) and thickness (m_0) of the upper confined layer. The greater m_0 and Ko, the better the natural protection of a confined aquifer. There is often no data on leakage coefficient. In this case, a lithological composition of this water confining layer is analyzed, which determines its filtration properties.

An important natural factor that determines protection properties of both confined and unconfined groundwater is the sorption capacity of water-bearing rocks and, particularly, the upper-lying ones.

In concluding this chapter, which has been devoted to one of the most important problems of present-day hydrogeology, namely protection of groundwater from pollution, I wish to express some important opinions.

1. An essential disadvantage of the above considered methodology for assessing groundwater's natural protection is the difficulty and uncertainty in determining the actual contribution of each factor (in other words their weight function) in

the formation of groundwater's natural protection properties. Here, an expert's approach is used. The available experience is based without specific quantitative assessments, and some factors are given a greater role (and correspondingly, a greater sum of marks), and others, a lesser one. This approach, characteristic of most methodologies using an assessment with marks, is of a qualitative character and gives the assessment results and protection (vulnerability) maps a somewhat conditional character. This comment relates to many presently popular methodologies, including DRASTIC, V. Goldberg's, etc. Thus, compiling maps of vulnerability of the above type should be considered the first very important and necessary stage for a quantitative assessment of groundwater vulnerability, which makes it possible to obtain an idea of the level of natural groundwater protection for a specified region.

2. The second stage, often almost more important, is in assessing the time period during which contamination (or contaminated water infiltrating from surface flow) reaches a groundwater level. This time period and the rate of pollutant migration into the aquifer can be estimated only through more-detailed investigations if there are data on sorption properties of the vadose zone and aquifer and on migration parameters. In this case, in calculating the time required for a pollutant to reach an aquifer, the available mathematical models and certain computer programs can be used, such as the latest modifications of Modflow, CHEMPATH, HydrogeoCHEM, etc.

   The use of such an approach for assessing groundwater protection (vulnerability) can be exemplified by A. Belousova (1994), who performed a regional assessment and mapping of groundwater vulnerability to radioactive pollution for some regions of the European part of Russia. Investigations were made for cesium-137 and strontium-90, and the time period was calculated during which these radionuclides will reach the groundwater table. Then this time period was compared with the half-life of these radionuclides. Territories where the calculated time period is less than a half-life (i.e., territories where groundwater is not protected from radionuclide contamination), more than a half-life by 1–2 times (relatively protected territories), and where this time period is more than the radionuclide half-life by more than twice (well protected territories). In general, the time period during which these radionuclides can reach an aquifer is directly proportional to the thickness of overlying deposits (or vadose-zone thickness), effective rock porosity, soil dry density and coefficient of sorption distribution of the radionuclides; and it is inversely proportional to the infiltration recharge (Belousova, 1994; Zektser, et al., 1995).

3. The degree of the above-considered natural factors' impact on groundwater vulnerability is different. Thus, it is quite obvious that for unconfined water, the main factors causing contaminant penetration from the surface into the groundwater are thickness, composition, and sorption capacity of rocks in the vadose zone. For confined water, the main factors are a ratio of levels (heads) in the assessed and overlying unconfined (i.e., potentially affected by contamination) aquifers, and filtration properties of the confining layer. Groundwater vulnerability must be determined separately for the main aquifers used now or in the future for water supply or irrigation. In these maps of vulnerability, quantitative characteristics are provided, and there are hydrogeological conditions given of the assessed territories. These are the primary main factors facilitating natural protection of groundwater from contamination.

4. It should be noted that, under disturbed conditions, compared to natural or already formed circumstances, the impact of the main factors on groundwater vulnerability can change. Thus, intensification of groundwater withdrawal from an unconfined aquifer causes an increase in groundwater depth to occurrence (the increase of vadose-zone thickness), which changes (in many cases improves) its protection from contamination from the surface.

   Intensification of water withdrawal from a confined aquifer is quite different. For instance, if under certain natural conditions there is actually no leakage of potentially contaminated groundwater from above into a confined aquifer, then under exploitation, when heads are decreasing in an assessed aquifer, a change of level ratio between confined and overlying unconfined aquifers will essentially increase the danger of polluting an aquifer and might even cause it to change into an unprotected one.

5. The danger of groundwater pollution depends not only on natural factors determining the extent of groundwater protection, but also on the general human impact on the environment, i.e., on the content of pollutants of anthropogenic origin in the environment. To estimate quantitatively the general pollution of the environment, it is advisable to use the technologic load modulus m (expressed in tons per year per sq. kilometer), being the ratio of the average amount of pollutants discharged annually to the area under investigation.

   To characterize the interconnection of groundwater pollution and environmental pollution, V.M. Goldberg suggested in 1987 introducing the index of groundwater susceptibility to pollution P, defined as the ratio of the technologic load modulus m to the groundwater protection extent index S, expressed in numbers, i.e.,

   $$P = mt/S,$$

   where t = time to accumulate a certain volume of pollutants in the territory under estimation, in years. The protection extent index S is determined from a sum of numbers characterizing the effect of a certain factor (see above). Thus, groundwater susceptibility to pollution is directly proportional to the human impact on the environment and is inversely proportional to the natural groundwater protection extent. The index P shows the interconnection between environmental pollution and the possibility of groundwater contamination. This index can also be represented in a groundwater-vulnerability map in different gradations (shadings).

6. Regional assessment and mapping of groundwater vulnerability is based only on the analysis and treatment of available hydrological and hydrogeological data without carrying out special expensive drilling and test-filtration or laboratory works. Therefore, there is no need for major financial expenditures.

Results of quantitative assessment and mapping groundwater vulnerability to pollution are of great practical value. They can be used in the following cases:

- to develop a strategy for groundwater use and protection in areas with differing natural vulnerability
- to verify the plans for placing and development of large industrial and agricultural projects that will generate hazardous wastes and wastewater

- to verify the groundwater use for water supply and irrigation and places for potable-water supply well fields, as well as to predict groundwater-quality changes under human impact
- to give a hydrogeological proving for different water-protecting measures
- to examine a choice of places for accumulating and storing wastes

One of the main practical results of assessing and mapping groundwater vulnerability is a possibility to compare different territories by groundwater protection from contamination and to decide what territory is better protected, and where there is a greater danger of polluting wells, pumping the groundwater for water supply, where there is a necessity of water protecting measures. Besides, these maps can be used by municipalities and other organizations for planning the measures for improving the ecological situation.

In conclusion, emphasize that groundwater vulnerability maps are of great practical value. These maps make it possible:

- to develop a strategy for groundwater in the area with different natural vulnerability
- to prove the plans of placing and development of large industrial and agricultural projects

The present-day assessment of groundwater vulnerability makes it possible to take up necessary measures for preventing negative changes in its quality and contamination. As a rule, Vrba notes (Vrba and Zaporozhets, 1994), if groundwater vulnerability is not assessed in time, and groundwater use and protection strategy are not decided upon, then the cost of polluted groundwater recovery will be much higher than the cost of measures for its protection. Assessing and mapping groundwater vulnerability must be an integral part of policy and strategy for use and protection of the groundwater in specific regions. These works must be closely correlated with a general current and perspective plans of environment protection.

7

Ecological Consequences of Groundwater Pumping

7.1 Influence of Groundwater Withdrawal on River Runoff

The change of interrelation between groundwater and surface water is the most significant ecological consequence of groundwater withdrawal in addition to its reserves depletion, its level decline, and cones of depression formation. The former should certainly be considered when assessing reserves and ecological consequences of coastal well fields operation of the so-called "infiltration type."

Coastal well fields are located along rivers and their safe yield is almost completely formed by the base flow of the river. It is evident that the functioning of these well fields immediately affects river runoff. Thus, it is understandable that numerous scientific investigations are devoted to studying interactions between surface water and groundwater under exploitation of the latter. Numerous international conferences and symposia are also devoted to these aspects.

The calculation of surface runoff affected by groundwater withdrawal was first considered by S. Theis, who analyzed the exploitation of a single well with a constant yield near a river, assuming that a stratum is homogeneous and infinite in its strike in the direction from the river. Most aspects of surface water and groundwater interconnections were first theoretically proved by M.S. Hantush, who indicated that, when calculating well-field yield, the effect of water-bearing layers should be considered as well as leakage from adjoining aquifers, thus demonstrating the irregularity of hydraulic relations between aquifers and rivers.

Later, in the works of E.L. Minkin and others it was shown that it is not possible to prove conclusively for specific prospective yields of coastal aquifers without a reliable assessment of surface runoff changes caused by groundwater exploitation.

River runoff changes resulting from groundwater exploitation are caused by a number of natural and anthropogenic factors. The most important are the following:

- character of hydraulic connection between the assessed aquifer and a river in different seasons determining the regime and dynamics of groundwater discharge into the river; this hydraulic connection character depends primarily on the relationships of aquifer levels and a river under both natural conditions and groundwater exploitation, regularity (completeness) of the river downcutting into the aquifer, and river bed transmissivity, determined by the degree of ground content and silt deposition in bed sediments
- annual and long-term seasonal variability of the river runoff
- character and amount of aquifer recharge and discharge, including a possible change of evaporation from the groundwater surface due to decline of its level under exploitation

- well-field yield and the distance between wells and the river bed
- filtration properties of the exploited aquifer

Consideration of these factors is necessary during a quantitative assessment of prospective coastal well-field yield and to determine possible river runoff changes caused by groundwater exploitation.

In most cases, river runoff changes occur under conditions of groundwater withdrawal resulting in hydraulic connections with rivers. However, the character, trend and scale of such changes can vary within a wide range at different stages of well-field operation. Because of the inertia of water-bearing systems, a change in river runoff will not manifest itself immediately. It is essentially smoothed out and lags in time. Moreover, in some cases river runoff can be even increased by discharge of used groundwater into the river. This groundwater is pumped out of deeper aquifers not drained by the river in the investigated territory, i.e., having no hydraulic connections with it. In these circumstances, namely inertia of the aquifer system, the probable discharge of already used groundwater into the river should be always considered when working out water management balances of separate regions and territories for a specified time period. Already developed techniques and models permit predicting possible groundwater exploitation with a high enough accuracy. This makes it possible to substantiate a system of managing coastal aquifer yield for preventing a catastrophic decrease of the river runoff or one that is unacceptable for different branches of economic activities (fisheries, navigation, and recreation). In some cases, such as a predicted significant decrease in low river runoff due to the impact of groundwater exploitation, special compensation measures like construction of regulating dams, etc., should be specified.

Depending on the relationships of different sources that contribute to groundwater safe yield of an exploited aquifer, the value of the possible river runoff decrease changes noticeably over time. Even during groundwater discharge into the rivers under natural conditions, surface water runoff can remain constant for a considerable time.

When assessing river runoff changes, hydrodynamic methods of calculation (including analytical, numerical, and analog, which are described in detail in special literature) are widely used. Therefore, only certain examples and results of separate assessments of intensive groundwater withdrawal effect on river runoff are given in this chapter.

A detailed analysis of the effect of groundwater exploitation on river runoff, given in detail in Chapter 2, was made by Yu. O. Zeegofer (1998).

Calculation of the change in separate water-balance components for certain river basins indicated that, due to groundwater withdrawal, inflow to the well fields from aquifers outside the Moscow region has increased. Here, groundwater discharge into rivers that amounted to 30% of the natural flow for the period of exploitation has decreased. However, the real loss for an average river runoff of 50%, even without considering return water discharge into the rivers, is about 10%. This often correlates with the accuracy of river runoff calculations.

Calculations indicate that a decrease of the river base flow, which drains only the upper rock masses in the fresh water zone and does not exceed 10%, will actually not affect the value of the mean annual river runoff. Rivers that drain the lower rock masses in the freshwater zone, and that are constituted by carbonate rocks of the Carboniferous Age — to which the main exploited aquifers are confined — are less affected. According to Yu. O. Zeegofer's calculations, river base flow out of these aquifers for the period of their intensive exploitation (60–80 years) has reduced by almost 50% below natural conditions. However, this refers to rivers, such as the Klyazma, the Moscow, and the Oka, etc., large enough to have a constant, considerable natural runoff that, by its absolute value, exceeds a decrease of groundwater inflow into these rivers by many times. Therefore, the values of base flow decrease in these rivers through the effects of groundwater exploitation are compared with mean annual river debits, the loss to river runoff appears to be quite insignificant. It should be indicated that this important conclusion relates only to the larger rivers and mean annual river debits. However in low-water periods of low-flow years, river runoff decrease resulting from intensive exploitation of drained aquifers can be very significant, and should be taken into account when considering different measures and constructions (hydrotechnical constructing, recreational constructions, factory developing, etc.) in rivers.

However, it should be kept in mind that, in river valley complexes, connections between confined and upper unconfined aquifers often change as there is no continuous confining bed between them (as, for example, in some river basins in the northwestern part of the Moscow region). Exploiting groundwater can then cause an activation of geodynamic processes in the coastal zone.

All the above, mentioned on possible river runoff losses due to groundwater exploitation, is true without taking into account return runoff, i.e., utilized groundwater discharge into the river. If it is considered that most of the utilized fresh groundwater is then again discharged into the river, then the loss to river runoff will actually be negligible. As utilized water discharge often occurs in a different place from that where groundwater was discharged into the river, a redistribution and change of runoff regime in separate places of the river, in comparison with a natural one, is possible. This phenomenon is complicated by the discharge of utilized groundwater. This also makes it difficult to analyze river runoff changes under the impact of groundwater withdrawal. An increase in debit caused by utilized groundwater discharge is observed in separate sections of rivers, particularly in the upper part of small rivers and springs.

A catastrophic decrease in river debits in places where large well fields are concentrated has been registered by English specialists (Owen, 1991).

Polish scientists give interesting data about the impact of long-term groundwater exploitation on the water regime of the River Drama basin (Kovalchik and Krolka, 1993). Here, one of the main tributaries' debits, which earlier drained an aquifer in Jriassic and at present recharges groundwater, considerably decreased under the influence of groundwater withdrawal by mine workings. Due to the formation of a deep and wide cone of depression, changes in both the basin hydrodynamic boundaries and river runoff general structure occurred in the exploited aquifer surface.

A substantial decrease of debits is marked in the Tokyo area that was caused by a decrease in the debit of springs or even their drying up, which was caused by intensified groundwater withdrawal (Arai Tadashi, 1990).

Intensification of groundwater withdrawal from the upper aquifers in the southeastern part of the coastal plain (USA) has caused the base flow of rivers and lakes to decrease in this area (Testa, 1991).

There are data (Zhorov, 1995) that, in some areas of Germany, indicate that mean annual runoff of small rivers and springs due to intensive exploitation of drained groundwater decreased by 30–40%, and in a low-water period some springs actually dried up.

In the area of Minsk (Belorussia) groundwater levels of a fluvio-glacial aquifer declined so much that spring runoff actually ceased, and river runoff in the low-water period considerably declined. This caused city authorities to impose limitations on groundwater withdrawal.

The given examples show the possible negative effect of considerable groundwater withdrawal on river runoff changes. Therefore, in practical hydrologic-hydrogeologic investigations, primarily those connected with water supply, special works for assessing the interaction between surface and groundwater in different periods of developing well fields are needed.

7.2 Influence of Groundwater Withdrawal on Vegetation and Agriculture

As noted in the previous chapters, in areas where an interconnection between groundwater of exploited aquifers and the upper unconfined groundwater is close enough, a decline of unconfined groundwater level caused by exploitation is observed. This groundwater level decline can affect the state of the landscape. Vegetation is the most sensitive element of landscapes, reacting almost immediately to changes in the groundwater table.

The effect of groundwater-level decline on vegetation is caused by the predominating regime of vegetation nourishment, namely, automorphic or hydromorphic.

Under automorphic regime of nourishment, vegetation roots do not reach groundwater level or the height of the capillary fringe, and get water only through infiltration of atmospheric precipitation into the root zone. The root-zone depth plus capillary-fringe height is often called the critical depth. If the depth of groundwater occurrence is more than the critical depth, a nourishment regime is automorphic. If groundwater occurrence depth is less than critical depth, groundwater activity participates in vegetation nourishment, and this regime is called hydromorphic.

The experience of numerous tests and experimental investigations (Zhorov, 1992; Zeegofer et al., 1998) indicates that the depth of root zone for most plants does not exceed 5 m. Thus, in Prioksk National Park near Moscow, the maximum depth of a pine-root zone does not exceed 3 m (in only one out of 17 cases it is 3.9 m), for an oak it is 5.1 m, for a linden 2.5 m, for a birch 3.4 m, and for an aspen 4.4 m. In this case, the main root mass is in the upper half-meter zone, i.e., in the zone of intensive infiltration of atmospheric precipitation. Similar results were obtained in the territory of Argedinsk National Park, where the main root mass of pines at the age of 10–15 years and 60–65 years was at depths of 0.4 and 0.8 m correspondingly (Sudnitsin, 1979). An analogous system in the root zone was observed for grassy and cultivated plants.

V.S. Kovalevsky (1994) indicates a parabolic type of dependence of ecosystem efficiency on groundwater levels. This parabola extremum corresponds to an optimal depth of groundwater level for an average vegetation period. According to different experimental studies, an optimal depth of groundwater occurrence for a cotton plant's vegetation period is 1.2–1.5 m, for most vegetables from 0.7–1.5 m, and for gardens 2–3 m on the average. Maximum output of coniferous forests widely spread in a humid zone of the East European Plain is observed in sandy-loam deposits when the groundwater level occurs at a depth of 1.2–2.0 m.

The height of the capillary fringe is caused by a lithological rock composition in the vadose zone. For sands of different granulometric composition, the rise is from 0.1–0.5 m, for light loams and peats up to 2.0–2.5 m, for heavy loams up to 3.0–4.0 m.

The above data demonstrate an important practical conclusion. When the level that has a hydaulic connection with a developed aquifer level is lower than the critical depth, then groundwater-level decline caused by water withdrawal will not affect vegetation in any way. In other words, it means that if the groundwater levels in sands are below 5 m and in loams below 7 m, then no amount of intensive groundwater exploitation will affect vegetation communities. However, this conclusion is true only for typical vegetation of platform structures in humid zones and zones of moderate humidity. For plants in a temperate climate, e.g., a eucalyptus, critical depths and, hence, water withdrawal impact on the character of plants, will be quite different. Unfortunately, data concerning groundwater exploitation on vegetation communities in the arid zone are absent at present.

Under a hydromorphic moisture regime, the best conditions for growing plants, both natural phytocenosis and cultivated plants, are observed at depths of groundwater levels from 0.5–2.0 m. The main root mass is in the same range of depths. Hence, a decline or a rise in the relative groundwater table can have negative effects.

Examples of negative groundwater withdrawal on vegetation are given below.

In the coastal zone of water courses in the Balaton basin mountainous areas, Hungarian scientists have noted the disappearance of some vegetation, which is caused by karst aquifer level decline due to intensive water pumping of mines and pits.

Studies made in 1996 indicate that in recent years, groundwater level decline in lowland sites has caused losses in the flora and fauna of national parks in the Netherlands. Measures were developed to reduce groundwater withdrawal from bottom sediments in Netherlands coastal zones by means of external water supply sources (Quack, 1990; Speets and Keeijberg, 1990).

A quite difficult situation connected with intensive groundwater withdrawal was observed in the south of Spain. Here, in the Guadalquivir River estuary, due to considerable groundwater pumping, the area of swamp mass was reduced from 200 000 he to 27 000 he. This has negatively affected the paths of migratory birds as they fly from Europe to Africa and back (Luke, 1992).

V.S. Kovalevsky (1994) notes that an increase in thickness of the vadose zone and the depth of the groundwater table, caused by groundwater exploitation in the Severky Donets River valley (Russia) resulted in loss and drying up of oxbow (meander) and backwater (tidal) lakes. Also, forests are drying up, particularly tree apexes (tops), which has resulted in a change of vegetation species composition. For the same reason, a cedar grove is drying in the Urals, as well as oaks near the Lebedinsk quarry of the Kursk magnetic anomaly, and gardens in the area of Krasnodar, North Caucasus, Russia.

Groundwater exploitation in river valleys of the arid zone has resulted in the formation of vast cones of depression that caused a loss of moisture-loving vegetation such as hydrophytes in many areas and has caused considerable oppression of phreatophytes. Significant damage was caused by a major groundwater withdrawal (about 800 l/sec) in the periodically dried up River Karagengir valley (central Kazakhstan). Here, according to the data of M.A. Khordikainen, drying and loss of vegetation was observed, and an abrupt decrease in transpiration was noted. A change of moisture regime as a result of groundwater level decline caused drying of meadow grasses in the river valley sites adjoining the well field, and consequently, river runoff abruptly decreased.

Hence, it should be noted that, in some cases, groundwater exploitation results in draining oversaturated lands and thus positively affects grass productivity and species composition in the flooded meadows. Thus, in separate sites in the Prisukhonsk lowland, artificially drained areas, because of groundwater pumping and melioration, hard sedge vegetation of low productivity was replaced by succulent meadow grasses of high quality and far greater productivity (Kovalevsky, 1995).

A similar picture was observed while analyzing the impact of groundwater level changes on forest productivity. Thus, additional draining of the land as a result of groundwater pumping can improve the quality of a forest in the oversaturated zone, and can make it worse — even cause its loss — in the arid zone.

Cases are known where groundwater withdrawal results in the drying and even loss of marshes. This causes the oppression of swamp fauna and flora, the disappearance of hydrophyte swamp vegetation, or a change in its species.

In the former FRG, in the middle mid-1970s, a period of essential groundwater withdrawal coincided with a series of dry years that caused a significant groundwater level decrease in vast territories, and, as a consequence, a change of landscape and vegetation.

In the same years, groundwater level decrease caused by water withdrawal in West Berlin negatively affected the vegetation.

Numerous examples of intensive groundwater withdrawal affecting landscapes and vegetation in different parts of Germany are given in a detailed analytical review by A.A. Zhorov's Groundwater and Environment. Some of the most illustrative are given below.

It should be said first that Germany is a country where complex field investigations and observations were carried out on a full scale, including experiments, modeling, predicting calculations, etc., for assessing the effect of intensive groundwater pumping on different components of the environment.

In the Furberg field territory, in Lower Saxony, the water regime was disturbed for 470 he of forests and 1400 he of meadows because of groundwater level decline due to exploitation. This amounts to 5.5% of the whole Furberg field territory. The dependence of carbonate-layer moisture content on groundwater table decline under water withdrawal is illustrated by graphs for 105 botanical testing grounds.

For the last 25 years of groundwater exploitation, the groundwater table the east of Furberg decreased by 4 m, and, as a result, a low-sedge swamp with 15 species of plants turned into a clover meadow.

In the Rhine River valley (Hessen overgrown river banks) in the period from 1964 to 1982 the lowest groundwater levels were registered. In this period, groundwater abstraction increase coincided with some dry years. As a result, land-surface subsidences caused damage to forestation and buildings. The total loss, including damage to buildings, highways, and railroads, losses to gardens, wastes for irrigation and planting in agriculture, and loss of forests was estimated to be equal to 16 million DM.

Detailed data on the connection between over-land ecosystems, and vegetation in particular, with the groundwater table being lowered by its withdrawal, are also given by A. Zhorov for other areas of Germany where long-term, special, and detailed observations of separate environmental components (soil moisture content, groundwater levels in natural and disturbed conditions, composition and state of vegetation communities, river runoff conditions, etc.) are carried out. The author notes that a mechanism for assessing a landscape's response to groundwater-level decline, and the risk of changes in landscapes because of water withdrawal was developed while projecting new well fields in the former FRG, where, over a period of 10 years, a working group called Ecology and Environment carried out some projects using a united technique. As a result of these investigations, methods and structure of stage-by-stage work while estimating an ecological hazard connected with water withdrawal was formulated. The technique is based on analyzing a water regime of a soil-vegetation layer and determining the role of ground moisture content in it. Special attention here was paid to studying the maximum height of the capillary fringe for different soil types, to determining a maximum field-moisture capacity and a root-zone depth depending on the lithology and density of rocks, to analyzing the impact of organic matter on the soil humidity.

As a result, schemes for assessing the response of agriculture and forestry areas to groundwater-level decline were elaborated and are now successfully used for assessing the impact of groundwater withdrawal on the environment and to prove different nature-protecting measures.

The main measures to prevent or minimize a negative impact of groundwater withdrawal on ecolandscapes and vegetation are as follows: management of well fields (in most cases decrease of pumping out), groundwater artificial recharge, surface runoff regulation, and using special hydroeconomic measures.

The ever-existing objective contradiction between what might be wished for and the necessity of withdrawing as much groundwater as possible, and the probable negative effect of such a withdrawal on different components of the environment, must be considered, based on the data of a complex ecological monitoring that includes the analysis of long-term monitoring of groundwater exploitation by large well fields. Such an approach will allow us to prove the scale and regime of rational water pumping, considering the requirements of environment protection.

7.3 Influence of Groundwater Withdrawal on Land-Surface Subsidence

As a result of groundwater movement, mass transport and redistribution of substances occurs immediately in the earth's crust. Dissolved-substances export by groundwater discharge is one of the most significant processes of chemical-element migration, which determines the scale of underground denudation. Quantitative assessment of underground chemical denudation is made by calculating the amount of dissolved solids exported by groundwater or the time of soil-surface subsidence to a certain depth (for instance to 1 m) due to dissolved solids export by the groundwater. Under natural conditions, this process is not immediately obvious. For instance, in the Baltic artesian basin area, about 30 t of dissolved solids annually exported by the groundwater in natural conditions from an area of 1 km^2 cause land-surface subsidence by just 0.008 mm per year. On the whole, land-surface subsidence in natural conditions due to groundwater discharge occurs slowly (less than by 1 m for 100 000 years), so gradually that it is not noticeable by man, and despite the significance of this process to the earth's geological development, no practical important negative consequences for everyday life are observed.

Land-surface subsidence due to groundwater oil and gas pumping is a different matter. Land-surface subsidence caused by intensive groundwater withdrawal is considered in detail below. As indicated earlier, it is one of the negative consequences of intensive groundwater exploitation.

It is known that, in areas of considerable groundwater withdrawal, vast piezometric-level declines are formed, usually called cones of depression, and often covering tens of square kilometers.

Groundwater piezometric-level decline and stratum pressure changes cause stress fluctuations in rocks, changes of groundwater-flow rates and sometimes direction, increasing suffusion, and karst-process intensity. Water-level decline causes land-surface subsidence under certain conditions and downfall formations under different ones. Most widely distributed are subsidences in the areas where groundwater occurs in easily permeable sandy-gravel rocks of low compressibility, interlayered with poorly permeable but well-compressed clay ones. Groundwater head decreased under pumping, which increases effective pressure on the rock skeleton and causes consolidation of compressed sediments, and, as a result, land-surface subsidence.

Depending on the character of deposits, rock consolidation can be either mainly elastic, recovering under level rising; or plastic, resulting in irreversible rearrangement of a granular structure of deposits that are actually not compressible. Pebble, gravel, and sandy rocks are poorly compressible, but their consolidation occurs quickly and is elastic by nature, i.e., under groundwater-level rise, rocks are essentially deconsolidated. The main reason for surface subsidence is connected with consolidation of poorly permeable clay deposits. Head decrease in aquifers creates a hydraulic gradient from overlaying clays and other poorly permeable rocks inside an aquifer to easily permeable rocks. A complete increase of pressure in a poorly permeable layer occurs first in pore water, and with water removal is gradually passed to a skeleton of poorly permeable rocks. Due to the low hydraulic conductivity of these rocks, vertical water movement during the next pore-pressure decrease occurs slowly.

Cases are not infrequent when karst-suffusion processes develop in carbonate rocks, enclosing fresh groundwater of good quality. Mechanisms of these processes in a simplified variant can be described as follows. Carbonate rocks due to disturbances in accumulating the sediments and effects of physical and chemical weathering are usually pierced to a considerable depth by numerous cavities and caverns of different sizes and configurations, and filled mainly with loose sediments. Under a lasting and intensive pumping of carbonate-deposit-confined water, a considerable increase of filtration rates occurs (by tens and hundreds of times). This results first in redistribution of a loose filler and then to its complete removal. The roof of already formed cavities cannot endure the load of the overlaying sandy-clayey deposits that are saturated with water, and this causes a slow land-surface subsidence.

Land-surface subsidence and cave-ins often bring about dangerous consequences. Thus, land-surface subsidence can cause a rise of groundwater level and swamping of territories, damage to motorways, railways, water pipes, and other communications, change of riverbed gradients, and deformation of industrial and civil constructions.

A.A. Konoplyantsev and E.N. Yartsev (1983) describe numerous cases of land subsidence and surface collapses that were caused by intensive groundwater exploitation.

Land-surface subsidence is widely spread in the USA. G.F. Polland (1981) indicates that subsidence levels change from 0.3 m in the area of Savannah, GA, to 9 m in the western part of the San-Joaquin Valley, CA. Subsidences exceeding 1 m are observed in Texas, Arizona, Nevada, and California. In California, the total of land-surface subsidence amounts to 17 000 km^2. The most significant change of environment caused by groundwater pumping was observed in the San-Joaquin Valley, where there is about 1.5 million he of irrigated land, half of which is subjected to subsidence. In this area, the decline in groundwater level that results from its intensive exploitation has amounted to some tens of meters, which caused rock subsidence in separate sites up to 9 m (Yokoyama et al., 1995). As a result of irregular subsidences in some places there occurred disturbances in the use of canals, water pipes, and well fields that required considerable financial expenditures.

In San Francisco, land surface subsided by 2.4 m, which necessitated the construction and systematic increasing of special dams to restrict an invasion of gulf water into the land. In the coastal valley of Los Angeles, land-surface subsidence at a rate of 0.7 m/year was observed during some decades. It was caused by the combined impact of anthropogenic factors such as oil, gas, and groundwater pumping, fresh groundwater pumping into the areas of seawater intrusion, and modern neotectonic activity (Testa, 1991).

In Mexico City, land surface-subsidence has amounted to 10.7 m for the past 70 years. As a result, buildings, water pipes, paved roads, and sewer systems have been damaged. Fine Arts Place in the city center dropped by more than 3 m below the surrounding street level. To stop the subsidences, surface water was supplied to the city and groundwater withdrawal was limited. Today, groundwater withdrawal has been reduced to 100 m^3/sec, with the population now totaling some 30 million people. At present, special measures are being worked out for managing groundwater withdrawal and water-resources usage (Durazo et al., 1989).

G. Lind in his brilliant book *Water and a City*, published by UNESCO in 1983, offers clear examples of close interactions among all the environment components, including general water resources, groundwater, land surface, etc. In particular, he gives a detailed description of geologic-hydrogeologic conditions in Venice, where a catastrophic submergence of the city into the sea caused by excessive groundwater withdrawal from the water-bearing layers and the effect of tides in the Adriatic Sea. In the period from 1952 to 1968, the mean rate of land-surface subsidence was 5–6 mm/year. But by 1975, due to a decrease in pumping (the number of operating well fields was reduced by 60%), the land-surface level had risen by 2 cm (Lind, 1984).

Similar examples are characteristic for many coastal towns that intensively exploit groundwater.

Many examples of the effect of intensive withdrawal of groundwater on land-surface subsidence can be given. Thus, according to the data of the Indonesian Technological Institute (1992), soil in Jakarta has subsided in some places by 0.5 m since 1994 as a result of intensive groundwater exploitation. This has caused seawater inflow that results in groundwater-level rise and, in some places, the destruction of building foundations.

It has been noted in a generalizing work (Barends, 1995), that, in Mexico, Japan, and the USA, there are at present more than 150 areas where intensive pumping of groundwater and mining of mineral deposits have caused land-surface subsidences amounting to 10 m.

Even intensification of groundwater withdrawal in some seasons, for instance, in summer for irrigation, can cause subsidence of loose ground (Sakai et al., 1996).

Land-surface subsidence in the territory of the alluvial plain Sagamegawa (Japan) was revealed in 1973. At the present time, a subsidence rate resulting from groundwater head decrease due to its pumping, amounts to 2 cm/year.

In Bangkok (Thailand), considerable land-surface subsidence, caused by groundwater exploitation has been observed since 1960. At present, there are about 1500 operating wells, pumping water out of confined sandy-gravel aquifers overlain by clays (Eddleston, 1996).

In Xian (China) due to groundwater intensive pumping out there are 0.5 m cracks formed on the land surface (Foster, 1995). Land-surface subsidences caused by groundwater withdrawal are also observed in Malaysia (Haryomo, 1995).

Many land-surface deformations caused by groundwater withdrawal are observed in California and Arizona, where subsidence cracks amount to 3.0–3.5 km.

In some areas, land-surface subsidences cover considerable areas. Thus, land-surface subsidence in Taiwan resulting from uncontrolled groundwater withdrawal covers an area of 250 km^2 and the maximum subsidence amounts to 2.5 m. Seawater intrusion is observed along coastal plains of the island. Measures have been adopted in some areas for decreasing groundwater withdrawal (Chian Min-Wu, 1992).

Generalization of data concerning land-surface subsidence in Japan caused by intensive groundwater withdrawal out of deep aquifers was made in 1990. In that country, the total area that subsided below sea level under the effect of groundwater withdrawal amounts to 1200 km^2. In Tokyo alone, land-surface subsidence amounted to 4.7 m in the first 75 years of the 20th century.

A water law passed in Japan is aimed at decreasing groundwater withdrawal and increasing surface-water use. At present, land-surface subsidence in some areas of Japan has been lessened by a more-rational groundwater withdrawal and by the increasing use of water from reservoirs (Yamamoto Sori, 1986).

In some areas, land-surface subsidence caused by intensive groundwater withdrawal has resulted in destruction of available civil and industrial buildings. Such cases are known, for example, in Calcutta (India), by land-surface subsidence at a rate of 1.4 mm/year caused widespread damage to buildings (Mishra, S.K. and Singh, R.P., 1993).

The danger of seawater intrusion, along with land-surface subsidence in many coastal regions, necessitates limiting of groundwater withdrawal. This danger is noted in the seacoast areas of California (Fio et al., 1995), in the Cape Verde Islands, along the coast of the Gulf of Mexico in the USA, in some cities of Italy and in many other areas. Seawater intrusion has brought about negative changes in the quality of groundwater on the Israel coast (Melloul et al., 1994; Hamberg, 1989).

In many coastal regions in England there is a problem of saline-water intrusion into the aquifers — on the eastern coast, in particular, there is seawater intrusion into the Cretaceous aquifer, which has been intensively exploited for potable and industrial needs. Specialists have recommended measures for stabilization of water withdrawal to prevent further seawater intrusion (Spink et al., 1990).

Intensive saline-water inflow into exploited aquifers is observed along the Queensland coast of Australia, where water withdrawal is almost twice the present infiltration recharge (Hillar, 1993).

One of the main — actually the only — way to control land-surface subsidence and seawater intrusion into aquifers, is to limit withdrawal of groundwater from exploited aquifers and to introduce careful monitoring of the groundwater level. Only the development of strict regulations for groundwater withdrawal can prevent or reduce negative consequences caused by well fields functioning in places where geologic-hydrogeologic conditions create land-surface subsidence or seawater intrusion.

The best example of preventing land-surface subsidence and marine-water intrusion is under way in Texas. In the city of Houston, land-surface subsidence reached 4 m in some areas during 40 years of intensive groundwater exploitation that resulted in flooding a considerable territory with seawater. In 1976, state authorities took up measures for limiting groundwater pumping in vulnerable regions and for increasing groundwater artificial recharge. The intensity of land-surface subsidence decreased considerably. It should be noted that Texas is the only region where planned long-term investigations (including stationary observations, modeling, and predictions) of the effects of intensive groundwater withdrawal on land-surface subsidence is promoted by the government.

Formation of vast cones of depression in areas where the Potomac-Magoti aquifer, NJ, has been exploited has caused seawater intrusion into the coastal zone that has twice compelled major consumers to reduce groundwater withdrawal (Kropp, Nocido, 1988).

In concluding this chapter, it should be noted that the problem of land-surface subsidence due to groundwater withdrawal (similar subsidences are developed under oil and gas pumping) draws the attention of many countries, where scientists' efforts are directed toward studying the characteristics manifested by these processes. The main problems to be dealt with are as follows:

- predicting development of this process in different geologic-hydrogeologic conditions
- proving of rational groundwater-exploitation regimes, particularly in areas subjected to subsidence and karst-suffusion processes
- elaborating on recommendations for preventing or reducing the negative consequences of excessive groundwater withdrawal

8

Groundwater Use and Public Health

The last quarter of the current century is noted for a distinct aggravation of problems based on the disparity between the growing needs of mankind and available natural resources. A full-fledged supply of potable water to the population is one such problem. The global scale of the issue was obvious as early as 1977, at the UN Conference on Water Resources. Analysis of the situation, which exists in many countries of the world and is characterized by potable-water deficiency in terms of quantity and quality, as well as by the mounting scope of incidence of disease among humans due to the adverse impact of the water factor, prompted attendees to proclaim the 1980s the Potable Water and Sanitation Decade. This was the decision of the 35th session of the UN General Assembly.

However, judging by the results of a series of subsequent major international fora that also dealt with the analysis of the situation in the field of water and environment, the problem is not only as acute as ever, but has been exacerbated. The processes of continental water-quality degradation stemming from powerful anthropogenic pressures continue to figure prominently in current environmental problems. The global nature of these processes has grown particularly acute in the second half of the 20th century, and is still regarded as crucial. Therefore, it is only fitting that groundwater should become increasingly important, because at times it features better-quality indices than those of the surface water sources. This gives a priority rating to groundwater used for drinking purposes.

At the same time, rapidly accumulating water chemistry data indicate that such views undergo a fundamental revision, especially in areas where, by virtue of various circumstances, the natural conditions of groundwater formation and maintaining its quality are disturbed. It should be noted, however, that the quality of domestic water is the subject of special concern for the monitoring services responsible for protecting the population's rights to a healthy environment.

8.1 Medical and Ecological Significance of the Water Factor

The medical–ecological aspects of the problem of groundwater development are now associated mainly with providing full-fledged and medically safe water for domestic use.

It seems worthwhile to discuss major principles used to determine the role of the water factor in shaping human environment and health. This will facilitate understanding of the medical–ecological approaches to the problem of groundwater use.

The impact of water resources on living conditions and health of the population is determined by:

- the extent of a sufficient and safe level of provided household-drinking and recreation-domestic water use

- existing sanitation facilities in human settlements
- development of recreation and curative zones
- the environmental impact of wildlife species dangerous to humans
- climatic conditions

The state of water resources also determines the potential for development of existing and the establishment of new human settlements with a view to agricultural and industrial development of all areas. These processes have a direct impact on the living conditions and health of humans, as they are sensitive to changes in the atmosphere and water resources, soils, vegetation, types of work, the level of nutrition and quality of food available to the population, and the nature of migratory flows (Elpiner et al. 1992).

The health of the population takes shape under the influence of external pathogenic causes and biological peculiarities of a particular human population, which, together, constitute a complex of medical–ecological factors. The theoretical and methodological basis of medical ecology comprises the environmental hygiene and general epidemiology of infectious (including parasitic) and non-infective diseases.

The knowledge accumulated in the course of studies in the above-listed fields indicates that water factor (including groundwater used for drinking-water supply) can strongly affect the character and incidence of infectious and non-infectious diseases, heredity, and the peculiarities of human organism development. This knowledge allows the development of measures designed to protect the population against the adverse impact of the water factor. This reasoning holds for groundwater as well.

Growing anthropogenic pollution of natural waters, which was typical of recent decades, brought about an intense development of studies aimed at the medical–ecological assessment of the quality of water in water sources and drinking water subject to the anthropogenic impact.

8.1.1 Epidemic Aspects

The role of water in transmitting a number of intestinal diseases (enteric fever, paratyphoids, dysentery, cholera, salmonollosis, viral hepatitis, and other, rarer, diseases) has been proven by long-term research over a century. Infectious diseases caused by pathogenic bacteria, viruses, and protozoa or parasitic agents are the most typical and widespread health-risk factors associated with drinking water (Guidelines, 1993).

Drinking water's pollution with infected communal and household sewage, or pollution at the source or in the waterworks system, is an established cause of numerous outbreaks of intestinal infections (On the state, 1996). The contemporary methods of epidemiological analysis that are used to determine the paths of intestinal infections' spread are quite informative. These are based on the identification of pathogenic agents (microorganisms) discovered in potable water and in the excreta of infected people. The pattern of the spread of a disease also has its peculiarities. Where there is a water-distribution network, the diseases are grouped along the course of infected-water movement. In the case of a decentralized water supply, the diseases are usually related to a specific water source, e.g., a well (On the state, 1996).

Numerous parasitic diseases and those confined to natural foci (malaria, opistorchiasis, diphyllobothriases, tularemia, leptospiroses, forest-spring encephalitis, etc.) are also closely associated with the water factor. With reference to the drinking water problem, lambliasis (its pathogen being the protozoan flagellant of the Lamblia genus) is among the most serious and capable of destroying human intestines and livers. According to the current epidemiological data, potable water is regarded as a major path via which the pathogene of this disease is communicated. The fact that some people are affected by this parasite is normally considered a sure sign of its presence in drinking water (Be'er 1996).

8.1.2 Toxicological Aspects

The spectrum of water pollutants is exceedingly wide. It embraces heavy metals, numerous microelements, toxic organic compounds, radioactive substances, etc. The spectrum of diseases caused by elevated concentrations of these substances in drinking water is also wide. These include the diseases of cardiovascular, excretory, digestive, nervous, and immune systems, allergies, defects of heredity and development, etc.

These notions are based on the data of extensive studies aimed primarily at the development of drinking-water quality standards.

The data cited below give an idea about the nature of impact of substances most frequently occurring in drinking water (including groundwater) on the state of public health (Rakhmanin et al., 1996; Manual, 1994).

This information pertains to both positive and adverse effect of the substances, depending on their biological impact and concentration. The daily requirements of the human organism for a particular chemical element is understood as the possibility of that element's consumption from various sources and, above all, from foodstuffs. However, it should be borne in mind that it is essential that a certain fraction of biologically valuable elements be supplied to an organism in an unbound form. On the other hand, the concentrations of biologically crucial substances should not exceed the maximum permissible levels for drinking water so that these do not acquire a biologically opposite nature. It is to be noted also that when the case in point is the harmful impact of a substance, the effect of prolonged exposure to its unusually high concentrations is invariably meant.

8.1.3 Inorganic Substances

Copper—Daily requirements are 2.0–3.0 g. Deficiency of copper leads to atherosclerosis of blood vessels and heart, anemia, hypercholesterinemia. An excess of copper is fraught with congenital diseases, changes in the water-and-salt and protein metabolism, blood redox reactions, disturbances of the ovarian-menstrual cycle, course of labor and lactation. MPC–1.0 mg/l.

Zinc—Daily requirements for adults are 2–3 mg, for children and pregnant women, 5–6 mg. Deficiency of zinc may result in congenital diseases (dwarfism), a change in the activity of redox-reaction enzymes, disturbances of the ovarian-menstrual cycle, the course of pregnancy, a reduced sense of taste and olfaction, and specific diseases of dermal integument. An excess of zinc leads to anemia, a change of the function of the central nervous system. At the population level, increased numbers of liver and cardiovascular diseases.

Fluorine—Physiological optimum is at 1.2 –1.5 mg/l (depending on geographic location). Deficiency of fluorine results in dental caries. An excess of fluorine leads to fluorosis (speckled dental enamel), polyneurites, hepatitis, sclerotic changes of the bones, arterial hypotension.

Manganese—Daily requirements are 1.5 mg. Deficiency of manganese slows down the rate of growth, disturbs fat metabolism. An excess of manganese results in anemia, and upsets the functional state of the central nervous system.

Cobalt—Daily requirements are 40–70 mg. A deficiency of cobalt leads to diseases in the blood system, changes in morphological composition of the blood, suppression of immune- and redox reactions. An excess of cobalt results in a disturbance of the functional state of the central nervous system and of the thyroid gland.

Selenium—Requirements for humans have not been established and, presumably, are in the order of several micrograms, depending on the level of E vitamin available in foodstuff. Deficiency of selenium encourages the "white-muscle disease" syndrome, at the population level higher infant mortality. An excess of selenium leads to rapid dental caries in children, and malignant tumors.

Aluminum—Has a neurotoxic effect. MPC–0.5 mg/l. In some studies, Alzheimer's disease has appeared to be associated with excess input of aluminum into the organism, in particular in drinking water.

Barium—Affects the cardiovascular and hematopoietic (leukemia) systems.

Boron—Produces disturbances of the carbohydrate metabolism, reduced activity of enzymes, irritation of the gastrointestinal tract; in men, reduction of the reproductive function; in women, a disturbance of the ovarian-menstrual cycle.

Cadmium—Increases the cardiovascular, kidney, carcinogenic illness rate, causes stillbirth and bone-tissue injuries, disturbs the ovarian-menstrual cycle, and the course of pregnancy and labor.

Molybdenum—High concentrations of molybdenum are associated with greater cardiovascular-disease rate, susceptibility to gout, endemic goiter, disturbed ovarian-menstrual cycle. MPC–0.25 mg/l.

Arsenic—Has a neurotoxic effect, causing skin injury, carcinogenic diseases. MPC–0.25 mg/l.

Sodium—Causes hypertension, abnormal muscular tension. MPC - 200.0 mg/l.

Nickel—Affects the heart, liver, causes oncologic diseases, keratites.

Nitrates and nitrites—Cause carcinoma of the stomach, hemopathy (methemoglobinemia).

Mercury—Disturbs sharply the functions of the kidneys, of the nervous system.

Lead—Affects kidneys, central nervous system, blood-making organs, causes cardiovascular diseases, avitaminoses C and B.

Strontium—Causes injury of the bone system (strontium-induced rickets).

Chromium—Disturbs the functions of the liver and kidneys.

Cyanides—Affect the nervous system, thyroid gland.

Dibromchlor-methane and tetrachlor-ethylene—have a mutagenic effect, carcinogens.

Iron—Causes irritation of skin and mucous membranes, allergic reactions, blood diseases.

Sulfates—Upset the functional state of the gastrointestinal tract, produce diarrhea, affect gastric juice acidity (hypoacidic states).

Chlorides—Affect the condition of the cardiovascular system (hypertension, hypertensive disease). MPC - 350 mg/l.

The adverse effect is known to be associated with both soft and hard water.

Generalization of recently obtained data reveals the pathogenic role of calcium and magnesium content of drinking water.

Calcium—Daily requirements 0.4 to 0.7 g (1.0–1.2 g for pregnant women and sucklings). Deficiency of calcium in drinking water results in a greater number of lethal outcomes of cardiovascular diseases, more-severe rickets, greater brittleness of the bones, disturbed function of the cardiac muscle and blood coagulation processes. Too much calcium leads to urolithiasis, disturbances of the water-and-salt metabolism, early calcification of children's bones, slower growth of the skeleton.

Magnesium—Daily requirements 220–260 mg. Deficiency of magnesium results in a more severe course and a greater number of lethal outcomes of cardiovascular diseases, neuromuscular and psychiatric symptoms, tachycardia and fibrillation of the cardiac muscle. An excess of magnesium may trigger the syndromes of respiratory paralyses and heart block, an irritation of the gastrointestinal tract in the presence of sulfates. MPC–20.0 mg/l of magnesium chlorate.

In recent works, correlation analysis was used to show that oncological diseases can be associated with the low concentration of water hardness salts.

8.1.4 Organic Components

Groundwater used for the drinking-water supply always contains certain amounts of natural organic substances. A broad spectrum of such substances is known. These include aromatic and humic substances, compounds containing carboxyl, carbonyl, and hydroxyl groups, heterocyclic compounds, carbohydrates, lipoids, bitumens, etc. However, the total natural organic matter content in water is usually small and ranges from a few mg/l to a few tens of mg/l (Krainov and Shvets 1997).

In the context of medical–ecological problems, two groups of substances are of great importance. These are humic substances and the products of mineralization of nitrogen-containing organic compounds (nitrates and nitrites). Humic acids have no detrimental effects except for an unwanted color that can be caused by their excess concentrations. However, chlorination of water containing natural humic substances and bromides brings about the formation of trihalomethanes. Of most importance in this group of substances are bromoform, dibromochloromethane, bromodichloromethane, and chloroform, which have distinct carcinogenic effects. Establishment of the effect of the formation of secondary toxic substances due to water chlorination changed both the hygienic assessment of the natural organic compounds contained in untreated water and the previously widespread notion that chlorine-based disinfection of water is harmless.

Application of strong oxidants (chlorine, ozone) to disinfect water containing natural organic compounds is associated with the formation of another toxic substance — formaldehyde.

As mentioned above, nitrates and nitrites can give rise to very dangerous diseases. Increased nitrate concentrations in drinking water are associated with blood diseases (formation of methaemoglobin, a perverted form of hemoglobin). The role of nitrites is less significant. However, in the human organism, nitrates and nitrites can transform into N-nitrosoamines — carcinogenic compounds.

The list of organic substances of anthropogenic origin capable of polluting water sources, including subsurface ones, is fairly large and includes several hundreds of compounds. These are chlorinated alkanes, ethylenes, benzenes, aromatic carbohydrates, pesticides, by-products of water disinfection, and a large number of other organic components such as products of organic synthesis and petrochemical processes as well as plasticizers, dissolvents, detergents, paints, etc.

Many of these substances can produce one or several toxic effects: carcinogenic, genotoxic, mutagenic, nephrotoxic (adverse effect on kidney), hepatotoxic (effect on liver). It should be noted that the concepts of the adverse effect of elevated concentrations of inorganic and organic substances in drinking water are based on extensive laboratory studies on animals aimed primarily at the development of drinking water quality standards. However, studies aimed at establishing a correlation between human diseases and some natural or anthropogenic component of drinking water take on ever-growing significance.

8.1.5 Standardization of Drinking Water Quality

The requirements for drinking water quality are being established at both international and national levels. Most countries develop their national standards on the basis of "Guidelines for drinking-water quality" (WHO), EU Drinking Water Instruction 80/778/EU, and Standards of the U.S. Environmental Protection Agency. According to the data presented during XVIII Congress on Water Supply (Copenhagen, 1991), by that time, WHO recommendations were accepted by countries with a total population of 500 million.

The development of standards is commonly based on the experimental–toxicological method intended for the establishment of maximum permissible concentrations (MPC), which was most clearly formulated in the Russian hygienic science (Manual, 1975; Krasovsky and Yegorova, 1990). The applied methodology implies studying the effect of different concentrations of the substance on the self-purifying capacity of water (establishment of the MPC based on toxicological index of hazard).

In the latter case, acute, subacute, and long chronic tests on laboratory animals are used to form the maximally possible toxicodynamic description of the substance. The parameters of its toxicity are being established with allowance for possible embrio- and genotoxic, mutagenic and carcinogenic effects; the ability of the substance to functional and material accumulation is determined along with the most vulnerable functional systems of the organism. Long-term, highly sensitive tests made on animals enable the establishment of the subliminal concentration of the substance in water and the obligatory or recommended standard. The lowest of the three established MPC values is taken as the standard.

In addition to the implementation of an experimental–toxicological approach, studies in ecology and epidemiology are being performed based on ecological and demographic data, and the materials of special epidemiological studies are used. Studies in this direction show a change in the concepts that small concentrations of some rated hazardous substances invariably have harmful effects. The present attitude of WHO toward this issue has become more prudent (Guidelines, 1993). This seems to be the reason for introducing the two-level standards by U.S. EPA (MCLG — maximum concentration level, and MCL – maximum concentration level) and the guidelines of the European Community (GL: guideline level and MAC: maximum admissible concentration). However, this implies specifying the admissible degree of risk under the conditions of application of available water-treatment technologies and means of water-quality control (National, 1991).

A rapid increase in the number of water pollutants requires the development of quick or calculation-based methods of establishing their admissible concentrations. These problems are closely related to the development of techniques for prediction and assessment of public health risk. The techniques suggested for solving these problems utilize either the available evidence of experimental toxicology or retrospective analysis of the correlation between the population disease rate and consumption of water with a certain composition (ecological–demographic method).

Purposeful sanitary–toxicological and ecological–epidemiological studies for the substantiation of the standards are useful in developing prophylactic measures aimed at prevention of the adverse effect of the water factor. To a large extent, these data also are utilized in the current system of establishing cause-and-effect relationships between the observed pathology and the adverse effect of the water factor.

The present drinking water quality standards should provide water epidemiological safety as well. For this purpose, the present international standards require a total absence of pathogenic bacteria, viruses, and the agents of parasitogenic diseases in drinking water. Analysis of the biological composition of drinking water is fairly complicated, and no rapid techniques are known to allow the detection of pathogenic organisms; therefore, organisms serving as sanitary indicators of biological pollution (E. coli) were selected. The presence of these organisms is an indication of the fecal pollution of drinking water; this pollution implies the presence of agents of infectious and parasitogenic diseases. The dynamics of development of this group of standards show the growing stringency of the requirements to the admissible level of E. coli concentration in the water of water sources and in drinking water. The latest edition of the WHO Guidelines on Drinking Water Quality (Guidelines, 1993) requires an utter absence of these microorganisms in 100 ml of water under study. An exception is the requirements for the treated drinking water in the supply nets of large water-supply systems, where the presence of E. coli is admitted in 5% out of samples taken within a 12-month period. Admitting that these requirements are difficult to meet, the Guidelines considers it obligatory to disinfect drinking water entering the supply net.

Comparative analysis of the international standards that have been introduced and improved by WHO since 1958 shows that the system of indices under control is developed through an increase in their number. The regulations improve mainly through the incorporation of many organic compounds associated with strengthening anthropogenic pressure on surface and subsurface waters (pesticides, products of organic synthesis, and petrochemical industry). The recent international regulation documents demonstrate an extremely negative attitude to the secondary products of interaction between strong oxidizers (used for water disinfection) with organic compounds of natural and anthropogenic origin.

8.2 Medical and Ecological Conditions of Groundwater Used for Drinking Water Supply

8.2.1 Water Quality Assessment

The complex environmental situation that has formed in many developed and developing countries, especially in transitional-economy countries, requires the development of a strategy of activity aimed at human health protection in specific hydrological and hydrogeological situations.

The existing international and national water-quality standards facilitate the solution of the problems arising in this case. However, economic aspects associated with the necessity of heavy capital investments in water-protection measures, reconstruction of existing and construction of new water-supply systems, improvement of water-treatment technologies, and the selection and use of additional safe-water sources, make it necessary for local authorities to draw up their own schemes for implementing the targets determined by standards. Substantiation of the relevant decisions requires medical–ecological assessment of the quality of water used to establish the role of qualitative and quantitative characteristics of water composition in the deterioration of the health of specific population groups. The detection of a specific hazardous pollutant in drinking water with the concentration exceeding the implemented standard is just an indication of its possible influence on public health. This shows that, in this situation, the water factor can be among the causes of an elevated incidence of a certain disease. This is a tentative level of indication of cause-and-effect relationships, however, it determines the direction of necessary studies aimed at a more accurate assessment of the water-factor effect (Elpiner, 1995).

The focus of this problem is the necessity of the human health risk assessment associated with the environmental quality and, in particular, with the quality of drinking water. The problems arising in this context are due to the lack of reliable data and knowledge necessary for independently assessing the risk factors. The development of the new discipline, ecological epidemiology, is directed toward solving these problems. This discipline integrates data accumulated in the fields of health protection, statistics, and natural sciences. WHO gives growing attention to this line of investigations whose results can be used to determine the top-priority problems concerning the protection measures and control of diseases, assessment of medical and prophylactic measures and selection of necessary technical and technological solutions (Guidelines, 1993).

Determination of the degree of admissible risk was found to be a necessary component in using different types of economic analysis like cost–benefit, risk–cost for establishing drinking-water-quality standards. It should be noted that the hypothesis that the effect of chemicals has no threshold provokes objections of experts (Elpiner, Vasiliev, 1983; Environmental, 1976).

The methods used by present-day ecological epidemiology are aimed at collecting data on the roles of external factors and protection (adaptation) mechanisms of the human organism in a specific environmental situation.

In this chapter we note that the approaches used by ecological epidemiology are based on detection of the similarities in and differences between the manifestations of a specific pathology in specific areas with the establishment of the difference between the hypothetical risk factors (Environment, 1996). It should be noted that, taking into account background conditions such as lifestyle (stresses, alcoholism, smoking, narcotics, etc.), environmental pollution, heredity, as well as medical and sanitary services, the quantitative assessment of the correlation between population health and water quality is a very complicated problem that sometimes fails to be solved by the mathematical methods of multifactor correlation analysis (Environment, 1996).

The methods used in ecological–epidemiological studies can be divided into two major groups:

1. methods of collecting data on the exposure and diseases of individuals
2. methods of studying the relationships between the amount of a substance exposed in the environmental sphere under study and the exposure time, on the one hand, and, on the other, the sick rate at the population level

The reliability of the data obtained depends on the size of the sample groups in the population under consideration (case-control, cohort studies), and the quality of the check group. Available experience in the field of ecological epidemiology indicates that performance of this task calls for rather complex and expensive research based on comparing the sick rate situation in the representative groups of the population, groups whose living conditions are very similar, differing only in the use of potable water of a dissimilar quality as far as the content of a particular chemical component is concerned. As it is difficult to find very similar case and control groups, confounding factors should be taken into account. The task becomes even more complicated when the effect of different concentrations of the same substance is established, as, in this case, the need arises for a considerable enlargement of the number of surveyed population groups, provided the aforesaid conditions are observed. In so doing, usually the data of official medical statistics have to be considerably supplemented with purpose-oriented clinical studies of the population.

Problems with obtaining sufficiently reliable ecological–epidemiological data for vast areas called for the development of methods that would allow the assessment of public health risk factors by using present-day sanitary–toxicological data as a basis for the establishment of the degree of hazard and the acceptability of different water sources. Thus, four hygienic classifications have been officially adopted in Russia:

1. Extremely hazardous: Standards for these substances are established in accordance with sanitary–toxicological estimates of their hazard; the MPC are of the order of a few hundredth of mg/l and less (e.g., phosphorus, benzo(a)pyrene, some compounds of mercury, tin, and lead).

2. Highly hazardous: These standards also are established in accordance with sanitary–toxicological estimates of their hazard; the MPC are of the order of hundredths and tenths of mg/l (e.g., thallium, cobalt, tungsten, hydrogen peroxide, some elementorganic, heterocyclic, and halogen-containing substances, the compounds of nitrogen, phosphorus, etc.).
3. Hazardous: In most cases, the relevant standards are based on organoleptic features and the implemented MPC values vary within a wide range from a few mg/l to hundredths mg/l (these are mainly organic substances, e.g., amines and their salts, aliphatic hydrocarbons, and others).
4. Moderately hazardous: The standards are based only on organoleptic features; the maximum permissible concentrations vary within a wide range, though commonly they amount to a few mg/l and several tenths mg/l (europium is an example of inorganic substances in this group; substances of this group can be found in any type of the organic substances under control, e.g., in the group of aliphatic esters and spirits, halogen-substituted thiophosphates, etc.).

8.2.2 Distribution of Water Bodies in Accordance with the Character of Pollution

Water objects are ranked in accordance with the extent to which the MPCs are exceeded in them (Plitman, 1989). Four degrees of pollution are specified for the substances for which toxicological standards have been established:

1. admissible (concentration close to the MPC)
2. moderate (concentration three times as high as the MPC)
3. high (ten times the MPC)
4. extremely high (100 times the MPC)

A combination of these two classifications makes it possible to evaluate the degree of discovered level of water source pollution and to see to what extent it can be used for drinking-water supply. For example, detection of substances falling under the first and second hazard categories in a water source with a "moderate" degree of pollution is likely to produce initial symptoms of intoxication in some part of the population. In case of a high degree of pollution with these substances, the symptoms of intoxication are more pronounced, and pathological effects typical of the discovered substances are at an appropriate stage of development (On the state, 1996). In other cases, the classification is based on the difference between the values of the same health index.

Four to five categories are commonly distinguished depending on the extent of variations in the demographic and sick rate indices within the given area as compared with the check area or the average values for a river basin or the country as a whole. For example, the medico–ecological situation in regions (or cities) is divided into five categories:

1. satisfactory
2. relatively strained
3. highly strained
4. critical or extreme
5. catastrophic or a situation of ecological disaster (Pinigin, 1993)

An important point is that the above approach is based on the studies that were used for ecological–epidemiological interpretations of the character of exceeding of water quality standards (Plitman, 1989).

Thus, the methodological basis of the present-day studies into the cause-and-effect relationships between the population sickness rate and water factor allows productive combining of ecological–epidemiological and experimental–toxicological approaches.

The practical effect of these approaches is illustrated by a wide experience accumulated by Russian scientists in their studies reviewed in the section dealing with research activity.

In the aforementioned system of proofs of the cause-and-effect relationships between human pathology and chemical pollution of the water, the role of the water factor in shaping up the sick rate of a particular disease is disclosed inadequately. The probability of sick-rate emergence or aggravation may also be due to other environmental factors or the "life-style," both of which may have a similar pathogenetic trend.

The procedure proposed to deal with this problem (Rakhmanin et al., 1996) envisages a comparison of the daily average uptake by the human organism of chemicals ingested with potable water and of their total daily average coming from the other spheres of environment: air, foodstuffs, etc. Here, too, special studies are required that call for an up-to-date analytical base, trained personnel, and financing.

However, even now, once we have the fundamental proofs of the impact of the water factor based on the earlier-cited approaches that require no special studies, one is free to take a sufficiently substantiated medical stand: Elimination of this impact by water protection and special water-treatment measures is to reduce the level of discovered pathology. At the same time, the advanced system of measures aimed at reducing the role of other factors (food, soil, air pollution, smoking, alcoholism, and drug addiction) is to minimize the level of discovered pathology.

8.2.3 International Experience

Analysis of more than 350 studies published during the last decades of the effect of drinking-water quality on the public health reveals a wide range of studies carried out in many countries (Figure 8.1). These publications can be divided into three main groups:

1. Studies establishing the presence of hazardous substances in drinking water within a certain area and suggesting the necessity of special-purpose medical–ecological studies.
2. Studies where the fact of water contamination is established and the health risk associated with its consumption is assessed based on the available data on the pathogenic effect of the detected substances (for example, with the help of data that have been used in the substantiation of the standards).
3. Studies applying methods of ecological epidemiology and toxicology to establish the cause-and-effect relationships between the population sick rate and the peculiarities of drinking water composition.

It is worth mentioning that most publications dealing with the chemical contamination of water belong to the first two groups. As for biological contamination, (either of a microbial or parasitogenic nature), in most cases, the relationship between the sick rate and the water factor is established by extensive epidemiological studies.

Notice that in the countries featuring environmental-protection agencies and high-level development of the municipal water supply systems, the role of water factor in the spread of infectious diseases still persists above all as a result of the use of contaminated groundwater from the topmost aquifer. Thus, according to the data of Global Consulting for Environmental Health (Moore et al., 1993), most (76%) of the 34 outbreaks of water-related infections recorded in 1991 and 1992 in 17 U.S. states were associated with the use of water

FIGURE 8.1
Geography of medical–ecological investigations of groundwater on human health. *Note*: Countries investigate are shaded.

from wells for drinking. Altogether, 17 464 persons were affected. In 11 outbreaks, the agent of infection was established, and in seven cases it was found to be Lamblia or cryptosporidia, pathogenic protozoan species. One of two other outbreaks was caused by water infection by dysenteric bacteria and the other was due to hepatitis A viruses. It is worth noting that the detection of groundwater infected with cryptosporidia is more and more frequently mentioned in publications. The fact is of great importance, because this agent causes an infectious disease that undermines the immune system of organism. Twelve outbreaks of this disease have been detected in the U.S. since 1985, with the agent of the disease detected in groundwater (Rose, 1997).

The use of poorly protected subsurface water sources can be a cause of outbreaks of intestinal infections of viral etiology. An example is the outbreak of acute gastroenteritis (up to 3000 affected persons) that occurred in Finland in 1994 because of use of water from wells infected by adenoviruses A and C, rotaviruses, and SRV viruses (Kukkula et al., 1997).

Water-related outbreaks of intestinal infections associated with the consumption of contaminated groundwater are mentioned in Russian official sources. However, these are mostly related to either wells or near-shore, river-infiltration water intakes that depend directly on the surface water quality (On the state, 1996).

We agree with authors who insist that greater attention should be paid to the protection of subsurface sources of non-centralized water supply. The up-to-date methods of disinfecting drinking water used in municipal water-supply systems notably reduce the risk of infection spread via water. They are especially effective in cases where protected aquifers are used. The cases of diseases associated with their use are caused commonly by secondary microbial infections of water in water-supply networks (Pozin, 1998).

However, no simplistic approach to the danger of biological contamination of subsurface water sources appears to be appropriate, especially under the condition of growing anthropogenic pressure. The evidence of this can be found in a vast database that disproves the widespread notion that aquifers are reliably protected against anthropogenic chemical pollution.

The most important evidence appears to be the data on relationships between the most widespread and dangerous cancers and consumption of chemically contaminated groundwater. The possibility of such relationships was first mentioned a few decades ago in papers where the character of population sick rate was compared for groups consuming water from subsurface and surface sources. A statistically significant increase in the number of cancer cases in a population group consuming surface water was detected by Kuzma et al. (1977) in Ohio, and Gottlieb et al. (1981) in Louisiana. Similar data were obtained in the studies conducted by Morin et al. (1985) in 473 American cities. We note that our results are close to those of earlier works of American, British, and Canadian scientists who have dealt with this problem.

However, studies conducted later in the U.S.A by the National Cancer Institute (Cantor, 1997) show an increased risk of the development of cancer pathology in the population groups consuming groundwater containing elevated concentrations of nitrites, asbestos-containing products, radionuclides, arsenic, and secondary products of water chlorination. At the same time, we find no indications of a similar effect of fluorides contained in water. However, Japanese researchers (Tohyama, 1996) established positive correlation between uterine cancer and the fluoride content of drinking water in 20 areas over their country.

Studies conducted by U.S. researchers in the Argentine confirm the correlation between an increased mortality due to bladder cancer and the presence of inorganic arsenic in drinking water (Hopenhayn-Rich et al., 1996). The importance of this problem for groundwater is noted in other papers (Haupert et al., 1996).

The risk of an increase in the incidence of cancer is associated with detection of carcinogenic organic compounds of anthropogenic origin in subsurface sources. The National Cancer Institute (Allen et al., 1997) draws particular attention to the pesticide contamination of groundwater. The authors suggest a correlation between the increased incidence of breast cancer in Hawaii and the consumption of groundwater containing chemicals such as chlordan, heptachlor, and 1,2-dibrom-3-chloropropane. Studies of Technical University of Denmark (Bro-Rasmussen, 1996) show that a number of pesticides (DDT, Lindane, dieldrine, and others) can persist in groundwater for a long time. Croatian researchers (Goimerac et al., 1996) detected contamination by the widely used carcinogenic herbicide atrazine. In an Egyptian region, pesticide contamination was detected in groundwater, foodstuffs, and soil, and later in women's breast milk (Dogdeim et al., 1996). Earlier works (Elpiner, Delitsin, 1991) reveal the mechanisms by which the pesticide reaches the breast milk via the chain: soil → drinking water → human being. A joint study of Israeli and Palestinian researchers (Richter, Safi, 1997) draws attention to the importance of taking measures aimed at an abrupt reduction in the adverse effect of pesticide contamination on human health.

Notice that all the above researchers agree that more comprehensive studies are necessary to assess the effect of pesticide contamination of groundwater on human health.

The health risk due to the presence of other organic compounds in groundwater is also the target of a number of studies. Of growing importance in this context are industrial-waste landfills, which contaminate soils and hence groundwater. Elevated toxicity of groundwater was detected at municipal landfills (Bruner et al., 1998; Najem et al. 1994). Accumulation of dichloroacetate (a by-product of production of industrial chemicals and medicinal preparations), which causes tumors and liver diseases, also can be associated with the effect of industrial-waste landfills. The toxicity of dichloroacetate was confirmed by experiments on animals, and the problem requires further toxicological studies (Stacpoole et al., 1998).

Of growing importance in assessing the health risk of groundwater are studies of the effect of a complex of contaminants. Among the recent works in this field we can point to the study by R.S. Young et al. (1995) from Colorado State University, which concerns the improvement of the methods of assessing the health risk of chemical mixtures.

Present publications give growing attention to the data on penetration of carcinogens into groundwater due to fuel leakage. American researchers reported on detection of methyl-3-buthyl ether (MTBE) in drinking water. According to B.R. Stern et al. (1997), nearly 5% of the U.S. population consume water with high concentrations of this substance (700 to 14 000 ppm). In the course of their studies of the risk of cancer due to groundwater contamination with MTBE, M.C. Dourison and S.P. Felter (1997) established that this compound remains toxic when it enters the human organism in drinking water. We believe that the effects of MTBE on cancer incidence require further investigations.

Recent studies by Taiwanese researchers yielded results that might cause anxiety. In their studies of a possible correlation between colon cancer and the level of hardness in drinking water in municipal water-supply systems, they correlated the frequency of deaths from this disease (1714 cases) with the number of deaths from other diseases (taking into account the hardness of water used by these people) and established a statistically reliable increase in the probability of colon cancer with a decrease in water hardness (Yang, Hung, 1998).

Based on a study conducted in 1997 in 98 cities and towns in Hyogo prefecture, Japanese researchers N. Sakamoto et al. also came to new conclusions concerning the pathogenic effect of water-hardness salts. They established a significant positive correlation between the death rate of cancer of the stomach with the Mg^{2+}/Ca^{2+} ratio in the water of subsurface sources and in tap water. Covariant analysis showed that Mg^{2+} significantly correlates with the death rate of cancer of the stomach. We are of the opinion that the relatively high Mg^{2+} concentrations in drinking water as compared with Ca^{2+} can be among the causes of stomach cancer in Japan. However, it should be mentioned that as early as 1980, B. Zemla (Poland) revealed a positive correlation between the low death rate of stomach cancer and high water hardness.

Studies of other toxic substances in drinking water showed a positive correlation of Al with Alzheimer's disease (McLachlan et al., 1996), As with bladder cancer (Hopenhayn-Rich et al., 1996), and NO_3 with gastric cancer (Yang et al., 1998).

A prominent example of solving the problem of revealing the specific effect of a group of chemical compounds is a study of the effect of secondary products of drinking-water chlorination (trihalomethanes that form when water poorly cleansed of organics is being treated by chlorine) on the cancer incidence in New Orleans (Environmental, 1976). A series of more-recent studies confirm the significant correlation between cancers and the trihalomethane content of drinking water (Hildesheim et al., 1998; Ijsselmuiden et al., 1992).

When considering possible adverse effects of the chemical composition of water of subsurface water sources, we cannot ignore other data on the role of natural components. For the past 20 years, experts of some developed countries have made attempts to essentially revise the traditional notion of water hardness as an index affecting its organic properties and the suitability for domestic needs. This was called forth by a series of studies aimed at the establishment of a correlation between drinking-water hardness and the incidence of cardiovascular diseases.

As early as the mid-1970s, the National Research Environmental Center, Cincinnati, used data on 135 cities in the U.S. (Environmental, 1979) to show a distinct tendency toward a decrease in the total number of cardiovascular diseases with growing hardness of drinking water consumed. Experiments conducted were based on the notion that cardiovascular diseases are caused primarily by some elements typical of both soft and hard waters rather than the water-salt composition. The U.S. Council on Environmental Quality evaluated the correlation between the hardness of water and 32 elements and substances that can be contained in it. The study established the presence (with a strong correlation) of α-radiation, ammonium nitrogen, arsenic, barium, beryllium, boron, cadmium, calcium, carbonate ion, chlorides, chromium, cobalt, magnesium, nitrates, potassium, phosphorus, silver, sodium, sulfates, and vanadium in hard water (with $CaCO_3$ content of more than 75 mg/l). No such correlation with hard water was detected for 11 other substances — copper, acids, fluorides, iron, lead, manganese, nickel, cyanides, salts of nitrous acids, phosphates, zinc, and β-radiation (Environmental, 1979).

It is worth mentioning that some theories about the protective effect of hard water on human organisms developed in the USA are based on the fact that the presence of a certain element in hard water accounts for the low mortality from cardiovascular diseases or their incidence in the population groups that consume this water.

The data were obtained by the National Laboratory in Oak Ridge in cooperation with the universities of Tennessee and Mississippi (Jackson) (Revis et al., 1980). These studies focused on the effect of groundwater composition on the development of hypertensive disease and atherosclerosis examined on biological models. Studying the long-term influence of drinking water with different levels of Na, K, Ca, Mg, Pb, and Cd content allowed the authors to establish the dominating role of Cd in combination with elevated concentrations of Ca and Mg in the formation of stable tendencies toward hypertension. The inverse relationship between the concentrations of these elements changed these tendencies to hypotonia. The role of other elements was insignificant.

At the same time, a significant reduction in the Ca content of drinking water with elevated Na concentration resulted in an increase in blood pressure relative to the check experiment where calcium content was at ordinary level. These experiments suggested the protective role of Ca in drinking water with respect to the hypertensive effect of other elements (Na and Cd). Moreover, soft water was shown to contribute to the development of atherosclerosis and an increase in the cholesterol concentration in blood plasma. The presence of Ca (100 mg/l) in water produced a protective effect with respect to Cd and Pb, which cause atherosclerosis. However, consumption of calcium-rich water (100 mg/l Ca) containing no other elements disturbed lipoprotein exchange and brought about the formation of atherosclerotic plaques in the aortas of experimental animals (Revis et al., 1980).

Further development of these studies with the use of ecological–epidemiological methods confirmed these observations and yielded some more-accurate and important data. Evidence of the correlation between coronary disease and the consumption of soft water (in particular, with Mg deficiency) was obtained also in Finland (Punsar, Karvonen, 1979), Italy (Masironi et al., 1980), Spain (Gimeno Ortiz et al., 1990), Germany (Sonneborn, Mandelkow, 1981), Russia (Plitman, 1989), the U.K. (Lacey, Shaper, 1984), Sweden (Rubenowits et al., 1996), Taiwan (Yang et al., 1996), and the Netherlands (Zielhuis, Haring, 1981) (Figure 8.2).

The published data show the significance of drinking-water composition as a whole rather than the specific role of each separate element. It should be noted, however, that the mechanism of influence of the above-considered elements on the conditions of the cardiovascular system are still to be studied. The observed effect can be a secondary manifestation associated with the impact of substances contained in the water on the nervous, regulatory, hormonal, and other systems of human organism. In any case, we have to agree with the principal conclusion concerning the existence of relationships between the conditions of cardiovascular system and the composition of drinking water consumed.

However, the lack of reliable and consistent data on the pathogenic role of water hardness has not allowed WHO to come to a final conclusion as to the implementation of the necessary standards (Guidelines, 1993).

8.2.4 The Russian Experience

In assessing the quality of groundwater used for drinking purposes in Russia, the role of anthropogenic pollution becomes increasingly important.

The data accumulated in Russia over the past few years serve as ample illustration of this (Chapter 5). We regret to admit that the environmental (in particular, hydroecological) sit-

FIGURE 8.2
Geography of medical–ecological investigations (impact of salts affecting water hardness) and secondary products of potable-water chlorination.

uation in this country has, in a number of instances, become catastrophic. Many of Russia's territories can be regarded as models for critical environmental phenomena taking shape. These are associated with a prolonged violation of the principles of observance of the admissible level of human interference in natural processes, neglect of the social and environmental interests of society, inadequacy of the current normative-legal, economic, and technological measures providing for the protection of the human environment, including water resources.

A study of the processes of water-quality degradation in a number of Russia's groundwater sources has revealed a broad range of reasons for the situation. The primary components of such a range are:

- injection of raw sewage into aquifers
- breakdowns at toxic-waste storages
- dumping toxic wastes in the rocks
- activity of virtually all oil and oil-products facilities
- violation of the sanitation-zone regime
- penetration of pollutants through the well heads or technically faulty casings

- introduction of substandard-quality water from adjoining idle aquifers or surface water courses and water bodies, including intrusion of sea water
- formation in groundwater of new and growing content of existing rated components resulting from the processes of physicochemical interaction in the "water–rock" system (Yazvin, Zektser, 1995).

At the same time, a relatively comprehensive system of regulations has been developed in Russia.

The requirements for the subsurface sources of public water supply are categorized in accordance with the source class. The State Standard (GOST 2761-84) "Sources of Public Water Supply" categorizes the subsurface sources into three classes:

1. Water quality meets the requirements for all indices.
2. Water quality deviates from first-class water in terms of some indices, but these deviations can be eliminated by aeration, filtering, or disinfecting, or the sources have unsteady yield that manifests itself in seasonal variations in solid residual within the limits specified by state standards for drinking water and requires preventive treatment.
3. Water quality can be improved to meet the requirements for drinking water with the use of treatment techniques specified in number 2 and additional techniques (filtering with presedimentation, application of reagents, etc.).

The quality of first-class water sources must meet all the requirements imposed on drinking water. The quality of second- and third-class water sources may be somewhat worse, because their development will incorporate application of treatment and disinfection techniques (Table 8.1).

In accordance with GOST 2761-84, the acceptability of a subsurface water source for municipal water supply must be estimated on the basis of analyses of the concentrations of chemical elements and natural composition indices (beryllium, boron, iron, manganese, copper, molybdenum, arsenic, nitrates, total hardness, oxidability, lead, selenium, hydrogen sulfide, strontium, sulfates, solid residue, free carbon dioxide, fluoride, chlorides, zinc) as well as the presence of industrial, agricultural, and domestic contaminants. The list of these indices should be approved by the local sanitary–epidemiological authorities, depending on the local sanitary conditions with both chemical and radioactive contaminants taken into account. The decisions should be made on the basis of a special study.

TABLE 8.1

Water Quality Indices of Subsurface Water Supply Sources

Index name	Water quality indices for classes		
	1	2	3
Turbidity, mg/dm^3, not more than	1.5	1.5	10.0
Color index, grad., not more than	20	20	50
pH	6--9	6--9	6--9
Iron (Fe), mg/dm^3, not more than	0.3	10	20
Manganese (Mn), mg/dm^3, not more than	0.1	1	2
Hydrogen sulfide (H2S), mg/dm^3, not more than	absent	3	10
Fluoride (F) , mg/dm^3, not more than	1.5--0.7*	1.5--0.7*	5
Oxidability (permanganate) mgO/dm^3, not more than	2	5	15
Number of bacteria of E. coli group in 1 dm^3, not more than	3	100	1000

*Depending on the climatic zone

It should be noted that in 1996 in Russia the new Sanitary Standards and Regulations of the control of drinking water quality were put into operation (SanPin 2.1.4.559-96). This progressive document adopts many requirements of the latest edition of "Guidelines for Drinking-Water Quality" (WHO, 1993).

Considering the limited barrier capability of water-supply systems, the applicability of water of a subsurface source in terms of its chemical composition indices is specified based on the list of MPC for water from water bodies, i.e., the Sanitary Standards and Regulations (SanPiN 4630-88) with allowance made for the possibility of improving the indices in the course of water treatment (decoloration, clarification, deferrization, etc.).

The data of Russia's Interdepartmental Commission for Environmental Security indicate that aquifers in many regions of the country are characterized by an unusually high content of iron, fluorine, bromine, boron, manganese, strontium and other rated microelements (On the state, 1996). Groundwater pollution by oil products is found to increase (Lujanchikov, 1996).

A closer study of groundwater pollution processes in the Moscow Region indicates that pollution is invariably preconditioned by the fact that most water-intake structures (on aggregate, providing 80% of the water produced in the region) are positioned near industrial sources of groundwater pollution. In some of the aquifers currently being used extensively, the standards of manganese, fluorine, chlorides, arsenic, selenium, and lead content are noted to be exceeded several times. One of the causes for this is a violation of the rules of livestock-manure storage and application (Elpiner et al., 1998). The cases of bacterial groundwater pollution are rather common, too.

Analysis of information gathered of late also indicates the necessity for a different look at the natural components of groundwater composition in the Moscow Region. For example, the groundwater enclosed in carbonaceous rocks has been found to contain an unusually high content (compared with the standard) of stable strontium. With some exceptions, its concentrations in the second and third aquifers from the surface reach 30–40 mg/l (Russia's current state standard of potable water quality sets the maximum permissible concentration of strontium at the level of 7 mg/l).

Also found within the Moscow Region are anomalous concentrations of fluorine in groundwater unaffected by technogenic pollution. Its concentrations vary from 0.2–0.5 mg/l up to 3–5 mg/l, exceeding the upper hygienic limit considerably (1.5 mg/l). Alongside strontium and fluorine, their geochemical satellites (barium and boron) are found in concentrations that exceed the existing standards.

The groundwater quality of recharge-type water-intake structures set up on river banks is directly related to both the level and nature of surface-water course pollution and the barrier functions of the filtering rocks. For a long time, during relatively moderate anthropogenic pressure on the surface water bodies, such intakes provided water of a fairly high quality.

Over the last 5 to 7 years, information of a different nature has been emerging increasingly. It is associated, above all, with an abrupt increase in anthropogenic pollution of the water bodies.

Toxic substances that virtually fail to be removed by the existing systems of drinking water treatment get into surface water courses in large quantities together with inadequately treated (or untreated altogether) sewage of different origin and with the surface runoff from urbanized and agricultural territories.

For example, every year, approximately 260 000 t of substances hazardous to human health are discharged into the Volga, out of this, 240 000 t via its tributaries the Oka and Kama rivers.

A substantial viral contamination of drinking-water-supply sources and of drinking water is established: in 1993, 1994, and 1995 the average percentage of tap-water contamination with enteric viruses was 1.6%; 1.4% and 1.28%, respectively; with hepatitis A virus antigen, 7.6%; 7.8% and 5.8%, respectively; with rotavirus antigen, 3.6%; 3.7% and 7.68%, respectively (On the state, 1996).

The above data underlines the importance of water quality at the recharge-type water intakes, and of the need to expand and improve control of the quality of the water obtained. The inadequacy of control already creates very unpleasant situations.

An example is the outbreak of acute intestinal diseases caused by bacterial contamination of water taken by an infiltration water intake near the Volga (On the state, 1996). This example is just one of the outbreaks of water-related intestinal infections. The number of water-related outbreaks has steadily increased during the past 5 years because of a high percentage of water-supply systems that fail to meet sanitary standards. The high level of bacterial and viral contamination of drinking water results in a permanently high level of incidence of acute intestinal infections and viral hepatitis A in some areas of the northern, east Siberian, and far east regions. From 1992 to 1994, the number of water-related outbreaks of intestinal infection in Russia increased considerably, totaling 62 with the number of affected persons some 9000. The use of bacterially contaminated water for drinking gave rise to an outbreak of cholera in Dagestan in 1994 (On the state, 1996). In 1995, 32 water-related outbreaks of dysentery were recorded in Russia with 4823 affected persons, while in 1992, only 16 such outbreaks were recorded with 1242 affected persons.

When considering the possible impact of surface-water microbial composition on the quality of water in recharge-type intakes, it seems necessary to take account of the latest results of scientific research. These indicate that, in view of the large-scale pollution of open water bodies and of the shifts in ecological equilibrium, aquatic microorganisms release resistant toxicants affecting the nervous, immune, and digestive systems in humans and lead to mutagenic consequences (Guidelines, 1993). There is no doubt that the possibility of these compounds penetrating the filtering rocks should be studied.

Over the last decade, a number of comprehensive surveys was carried out in Russia to study the cause-and-effect relationships between somatic (other than infection-induced) disease rate of the population and anthropogenic pollution of drinking water (On the state, 1996; Plitman, 1989; Rachmanin et al., 1996; Semenov et al., 1994, etc.).

These studies have made it possible to establish that a higher rate of chronic nephritis and hepatitis, a higher mortality, gestational toxicoses, and congenital anomalies of development in children are associated with the use of drinking water polluted with nitrogen-containing and chlororganic compounds (the towns of Kemerovo and Yurga). The use of groundwater with an "extremely high" natural content of boron and bromine has resulted in the growing sick rate of children's digestive organs in the town of Shadrinsk, Kurgan Oblast. Drinking water containing aluminum in concentrations five times the established standard had an oppressive effect on children's central nervous systems and immune systems in a settlement of Novgorod Oblast. When the population of the Middle Volga Region began to use groundwater with a hardness of more than 10 mg eq/l, which also contained 300–500 mg/l of calcium, the frequency of urolithiasis was found to increase. The same sources indicate that nearly 5% of the population of the steppe zones of European Russia and some areas of the Volga Region must use groundwater for drinking purposes with a high content of chlorides and sulfates, 3–5 times the existing the standard, which preconditions a high sick rate of cardiovascular diseases, chole- and urolithiasis. At the same time, cardiovascular diseases figure prominently among the causes for mortality of Russia's population (Prokhorov, Revitch, 1992). The sick rate of Lipetsk town-dwellers illustrates the relationship between an unusually high level of nitrites in drinking water and the suppressed blood-producing function of a human organism. Also, a direct relationship has been established between (a) the extremely high sick rate of the digestive and central nervous systems and carcinogenic diseases among the population of certain districts in the

Republic of Buryaria and (b) lack of some microelements in drinking water. An in-depth medical survey of the population in Krasnoyarsk Kraj and Amur Oblast has revealed an adverse effect on mineral metabolism and the functional state of the central nervous system of water deficient in the salts of calcium and magnesium.

Analysis of the data regarding the content of fluorine in drinking water used in Russia indicates that more than 60% of its residents fail to get this microelement in an adequate quantity. Sanitary-control bodies link this phenomenon with the extremely high level of occurrence of dental caries. In some oblasts (administrative subdivisions), 80% of children are affected by this disease (Elpiner, 1995).

The problem of secondary products of water chlorination having a carcinogenic effect has proved rather topical for Russia. The available data indicate that drinking water in half of Russia's towns does not meet hygienic requirements so far as the content of chloroform, a haloform compound, is concerned (Rakhmanin et al., 1996). The main reason is the mounting rate of water-sources pollution and frequent unavoidable hyperchlorination of water. The use of this technique meant that, in 1995, 7.9% of water samples taken from the waterworks of Nizhni Novgorod contained chloroform in concentrations that exceeded MPC (Rakhmanin et al., 1996). It should be remembered that chloroform is an indicator of the presence in the water of trihalomethanes, secondary products of chlorination that are active carcinogens. At the same time, the growing cancer rate among the population of Russia's numerous regions makes the problem of carcinogenesis just as vital as that of cardiovascular pathology (Prokhorov, Revitch, 1996).

As applied to the problem of groundwater use for drinking purposes, the emergence of secondary products of water chlorination remains as vital as ever because these are formed as a result of chlorine interaction with natural humic compounds or organic pollution. The former may be discovered in groundwater whose quality is formed under the influence of raised water-logged areas that are likely to occur in recharge-type intakes in connection with river water pollution, especially water polluted with chlorinated organics. However, it is not uncommon for river water to contain natural humic compounds, too.

8.2.5 Water Sources Choice

The above-considered data clearly demonstrate the importance of the present-day medical–ecological interpretations of hydrochemical data that must be taken into account in assessment of the conditions of drinking water use and in the selection of new groundwater sources.

The reason for this is the data on the ever-growing number of cases of contamination of subsurface sources of water intended for drinking as well as new data on the possible adverse effects of their natural composition.

To be able to describe the quality of potable water obtained from underground sources, it is essential that such water should belong to one of three groups of regions (Yazvin, Zektser, 1995):

1. Areas where fresh water of the underground aquifers, in terms of macro- and micro-component composition in natural conditions fully meets the drinking water requirements.
2. Areas where the content of microcomponents in fresh groundwater of particular aquifers exceeds the established standards. Case in point is hydrochemical provinces whose groundwater features heavier (relative to the established standards) concentrations of specific rated components. The hydrogeochemical provinces are characterized by a high level of background concentrations of specific micro-

components that approach or exceed the admissible level as well as a high (> 50%) frequency of the occurrence of concentrations of these components exceeding the admissible level of their content in groundwater.

3. Areas virtually devoid of fresh groundwater, where groundwater of a high mineral content is common, with a high content of sulfates and chlorides as well as characterized by a high total hardness. These areas adjoin those where the thickness of freshwater aquifers is insignificant, which accounts for the great importance of mineralized water leakage when these aquifers are used.

This class should be enlarged to include areas where groundwater contains anthropogenic chemical contaminants with concentrations exceeding the imposed standards. These are most commonly highly urbanized and industrially developed areas with insufficient natural protection of groundwater.

The natural conditions under which the quality of groundwater is formed undoubtedly affect the medical–ecological appraisal of its quality. Here, the impact of macro- and microcomponent composition must be distinguished. The former, indicating the hydrochemical category of groundwater, is essential for evaluating its organoleptic properties (the most favorable is the water of a hydrocarbonaceous category, calcium-manganese group, featuring moderate mineralization up to 500 mg/l) and for a probable pathogenic impact, where the content of macrocomponents, such as sulfates and chlorides, and hardness salts is greater than is prescribed by the norms. For example, there is an established relationship between urolithiasis and drinking water whose hardness exceeds 15–20 mgeq (Manual, 1975).

A large body of research papers makes it possible to pass judgment about the pathogenic importance of low-salinity soft water. They tend to associate a protracted use of such water with a cardiovascular pathology. This is blamed, among other things, on the deficiency of calcium and magnesium, which play the role of protectors in respect of toxic microelements frequently occurring in such water (Elpiner, 1995). In the data above we should draw attention to the possible influence of Mg-deficiency on cancer incidence.

The natural composition of groundwater in terms of microelements also makes it possible to evaluate its quality from the viewpoint of health safety when it is used as potable water. Here, it is important to know the category of hazard and level of discovered concentrations of the substances (in this case, microelements) present in the water. This group of substances may include boron, niobium, lithium, tungsten, cobalt, silver, bromine, etc., which belong to the second category of hazard. However, microelements may also include fluorine, whose lack or excess in a fluorine-treated drinking water may increase the dental sick rate. Despite a rather heated discussion of the need for drinking-water treatment with fluorine, WHO regards this technological method of water treatment as absolutely justified from the standpoint of medicine (Guidelines, 1993). So far as the provision of a minimum required content of other microelements in potable water is concerned (most of these augur well for a human organism in physiologically justified micro-quantities), the contemporary international and national standards for drinking-water quality ignore the issue because of inadequate experimental and epidemiological substantiation (Guidelines, 1993).

Materials cited in this chapter stress the value of medicobiological aspects of the problem of groundwater protection and use for drinking purposes. At a time of burgeoning industrialization and urbanization, groundwater essentially retains many of the advantages of clear drinking-water sources. However, increasingly negative examples of adverse, particularly man-induced, impact on groundwater quality are being accumulated. The bank of medical–ecological data currently taking shape indicates there obviously exists a direct cause-and-effect relationship between the sick rate of the population and the degrading water quality at the water sources reserved for drinking purposes, which include groundwater.

The new environmental situation calls for new approaches to the assessment of groundwater quality and intensification of measures aimed at groundwater protection and safe use.

This problem should be considered in the context of incorporation of environmental priorities into water-management activity. It should be realized that the problems arising in this field can be solved only through an interdisciplinary approach. The implementation of this approach is crucial for the full-scale development of medical–ecological studies with hydrogeological orientation. This aim can be achieved only based on the up-to-date hygienic, environmental, hydrogeological, social–demographic, and medical–geographic approaches. However, the most important in this context is the combination of medical investigations with comprehensive hydrochemical data. This problem can be solved only by joint efforts of experts in all the fields involved and on the basis of data exchange concerning the targets and progress of the investigation.

It is clear that the medical–ecological studies are expected to provide a reliable basis for the selection of new, safe, subsurface water sources as well as for the improvement of groundwater protection and water-treatment technologies. In other words, they are intended to support efficient water-management decisions allowing for human health protection priorities. At the same time, the lack of reliable data on the medical–ecological situation related to the hydrogeological conditions of water use as well as predictions of its evolution increases the probability of making incorrect decisions.

This circumstance calls for urgent development of the theoretical basis and techniques of medical–ecological zoning in terms of the relationships between the public health and the quality of groundwater that is either used or planned to use (Pinigin et al., 1998).

Another, just as important line of investigations is the development of methods of forecasting the variations in medical–ecological conditions, which also depends on the hydrogeological conditions of water use.

The development of the above-described approaches requires further improvement of the methods of establishing cause-and-effect relationships between the public health and the water factor. The research problems involved are associated with the elaboration on unified methods of ecological epidemiology and ecological toxicology. These problems obviously call for consolidation of the efforts of scientists all over the world. This is of great importance, because the problem concerns securing a safe drinking-water supply for the present and future generations.

9

The Impact of Human Activity on Groundwater Resources

9.1 Economic Activity and Groundwater: Common Aspects

A high tempo of anthropogenic activities, a huge scale of land development, and increasing intensity of the exploitation of vast areas causes intense transformation of engineering hydrogeological conditions and the geological environment as a whole. As a result, area water balance changes considerably due to human economic activities, i.e., new recharge and discharge items appear and their relationships in time and space change.

An additional amount of water flows into developed areas via water conduits, irrigation, etc. Land surface covered with different types of constructions interrupts the evaporation process. Buildings on embankments and the filling of small rivers and ravines have made groundwater discharge difficult, new water-bearing horizons are formed, and the functioning of different water-level-lowering systems and anthropogenic sources of recharge form an artificial-groundwater regime.

Human economic activities in the 20th century became the most powerful influence on the environment. Groundwater, being one of the most movable lithosphere components, reacts to this impact most directly. As a result, it is difficult to predict situations that are hazardous for man, and the results of his activities often appear. All this could not but affect the environment. A certain total conflict situation occurs. Mankind cannot exist without "using" the environment. At the same time, the environment is the habitat of man and requires constant preservation, restoration and, in most cases, regeneration. And all this occurs against the background of the earth's total territorial development and often including different global changes, for instance, climatic.

In a general case, it is reasonable to consider a system's "groundwater regime–anthropogenic activities" within a framework of special natural-technical and social system (NTSS), where corresponding subsystems are in special interconnections and interrelations that are often conflicting in nature. This is the basis for destabilizing factors to occur. This constantly leads to the formation of dangers for man while supplying him with his necessary groundwater resources. This can at any moment transform into an extraordinary situation with a certain degree of probability. At the same time, behavior of such an NTSS is difficult to predict. That is a source for uncertainties with which a decision maker will be faced in the future. In a built-up area it is difficult to foresee the character of new development, the location of water-supplying communication, changes in hydrogeological conditions under changing anthropogenic loads leading to unfavorable consequences, changes of groundwater regime, and protective measures to be taken up, etc. Hence, any decision-making activities for groundwater use are risky, as they must overcome these uncertainties.

Risk in this case is, in essence, a certain cost for ignorance of the main regularities in the groundwater formation in an area, insufficient knowledge of specific hydrogeological conditions, inadequate information on the engineering-hydrogeological situation, and misunderstanding of dynamically developing interrelations and interactions between NTSS elements. Finally, it is a cost for incompetence in overcoming uncertainties occurring in developed areas using groundwater resources in the process of human activities.

The uncertainties mentioned can be overcome by creative use of man's experience and the results of groundwater monitoring.

However, some uncertain elements always remain. Hence, the risk of adopting one or another hydrogeological decision for assessing groundwater resources under intensive anthropogenic activities is actually inevitable.

In the developed areas (urbanized, improved, located in a reservoir or storage backing zone and in a drainage zone) the formation of groundwater resources is essentially affected by anthropogenic factors, and change notably at different stages of development. Actually, great volumes of water are supplied to the developed areas. However, water's considerable withdrawal is also realized by different systems of well fields.

These two diametrically opposed processes are usually considered quantitatively: increase or decrease of debit in pumping wells.

Evidently it is not sufficient, as in the areas subjected to one or another type of anthropogenic activities, that groundwater quality changes considerably. Chemical, thermal, bacteriological, radioactive, and gas pollution limit its use. Vice versa, using different methods for groundwater purification (treatment), decreasing the intensity of its contamination, and also natural processes of groundwater self-purifying, lead to increasing these resources, i.e., to their recharge.

Thus, peculiarities of anthropogenic activities' impact on areas urbanized, improved, and affected by hydrotechnical constructions (reservoirs, big channels, etc.), should be considered.

A scale of anthropogenic impact on the groundwater becomes regional. Thus, predictive assessment of possible unfavorable human consequences should be made here. Primarily, a retrospective assessment of hydrogeological conditions interconnected with anthropogenic activities at that period is needed. Peculiarities of hydrogeological conditions in developed areas are primarily the result of constantly changing forms of anthropogenic activities here.

The next fundamental step is predicting possible development and change of anthropogenic activities, and, hence, the probable effect (scenario) of loads on the groundwater. This scenario, in its turn, undoubtedly depends on the character and dynamics of interacting natural, social, and technical subsystems in every specific NTSS, which are far from being identical.

It is necessary, then, to make predictive assessment of how new situations affect the formation of new impacts. To determine whether fluctuations of controlled indications fall outside a permissible, in these specific conditions, range of their possible deviations. It is probable by nature, in most cases.

It should be considered that anthropogenic impact on the object can be direct or indirect, i.e., to act on the object through a number of intermediate processes or actions. For instance, water infiltration caused by irrigated farming can make groundwater levels rise and adjoining, often built-up areas, may be swamped and water logged. Here, dangerous engineering-geological processes are provoked and buildings and constructions are damaged, etc. Hydrogeological predictions are objectively accompanied by a number of errors that provide for forming uncertainties, appearing to be due to our water-management activities and the decisions made.

There are three main groups of errors singled out:

1. methodical
2. measurement (under determining calculated parameters)
3. calculation (errors during organization activities are not considered)

Methodical errors are primarily caused by defects in systematization of hydrogeological and technical initial data, their typification, and choosing a calculated scheme for a well-field functioning (based on boundary conditions), as well as the level of adequacy of accepted engineering-hydrogeological model to a real situation. It should be considered that a hydrogeological object of investigation is unique and hence, a creative approach is needed in every specific case. As a result, a certain subjectivity is inevitable when elaborating a calculation scheme and a mere assessment method. It should be noted that methodical errors often appear with inadequacy of the model scale (including time) to prediction.

Measurement errors are errors in determining calculated parameters and hydrodynamic boundaries. These are errors of determining hydrogeological characteristics and parameters of available and potential disturbance sources. In the first case, errors can appear while measurements are being taken. In the second case, errors depend on the level of filtration heterogeneity of the rock mass, i.e., on the complexity of engineering-hydrogeological conditions and variability of the anthropogenic impact. They are connected by a procedure of spreading the available data into certain sites of the area, i.e., to their averaging. This procedure is statistically uncertain in itself. Inadequacy of available initial information to a scale of research that can be decisive under delineating boundaries of sectionally homogeneous studied area and under the following choice of boundary conditions should also be considered.

Calculation errors are the errors appearing immediately while calculations are being made. They are unavoidable, but easily removed.

Errors in different groups considered by us are incommensurable between themselves. Particularly serious errors in predictions are caused by mistakes in the first group. Thus, in most cases, it is unreasonable to make the requirements to test filtration works more strictly (to increase volume, accuracy, duration, and cost), leading to a decrease in the number of errors in the second and third groups. Therefore, prediction must be preceded by analyzing possible and admissible errors in the initial information.

All this indicates how difficult the prediction of an engineering-geological situation is on the one hand and how flexible and adaptable, on the other hand, an engineering protective system based on a "capricious" prediction that mostly depends on the author and his experience, must be.

We often make decisions without considering the ways of hydrogeological condition formation in developed areas, their changes in the future, and unfavorable consequences as a result. In this case, the unfavorable consequences of proposed protective measures must be kept in mind. Thus, the main purpose of a researcher is to determine what must be done at present to avoid possible unfavorable consequences for groundwater, environment, and people in the future.

During the current period of anthropogenic development, when numerous potentially hazardous uncertainties must be overcome, permanent (not long-term) predictions of possible engineering-hydrogeological situations based on a complex monitoring (groundwater plus anthropogenic loads plus socio-economy) are primarily needed when making management decisions. A permanent-situation engineering-hydrogeologic analog is needed. Systematic, adaptable regulations (a system of engineering and legal measures) of constantly originating conflict engineering-hydrogeologic (including ecological and social) situations in various NTSS is also needed. But this should be done within a framework of general strategy elaborated for this NTSS functioning.

Let us consider a general system of groundwater regime changes in developed areas. Most importantly, it is necessary to clearly think through the general situation and systematize the main factors forming the regime. They should be considered depending on the size of the affected area (regional or local) on groundwater recharge and discharge conditions recharge or withdrawal, on genesis (natural or artificial), on activity affecting the formation of a hydrodynamic situation (active and passive), and on the character of action (occasional and determined). The action can differ in time (systematic, periodic, and occasional). It should be marked that not all the factors can be traced, which is why observed and nonobserved factors, as well as controlled and noncontrolled, are singled out.

Regional factors (relative to an area considered or an object singled out) affect groundwater recharge or withdrawal and the corresponding rise or decline of its level. In the first case, it is groundwater damming from water reservoirs, irrigation systems, big channels, large technological storage, industrial objects consuming water in large amounts, infiltration of leakage out of massive collectors, and water filtration out of zones with washed up and filled grounds accumulating groundwater, etc. In the second case, the factors affecting groundwater recharge or withdrawal are formation of cones of depression due to functioning of large well-fields, systems of draining mine fields, large quarries, subway tunnels, swamps, water-level decline in rivers due to their deepening and dredging, etc.

Local factors also cause groundwater replenishment or withdrawal and consequently a rise or decline in its level. In the first case, it is a head, resulting from the damming effect of foundations dug for buildings and construction, from dams, filled-up ravines and gullies, flooded grounds where water accumulates, infiltration of leakages from water conduits and surface runoff due to its disturbances or insufficiently developed system of rain channelization (including periods of catastrophic atmospheric precipitation), water accumulating in grounds beneath trenches and ditches. In the second case, it is formation of cones of depression due to single wells and drains.

Due to the impact of regime-forming factors a radical change of water regime, often causing unfavorable consequences, occurs during development of the area and its consequent exploitation. A general scheme of groundwater regime changes due to anthropogenic impact and their consequences in developed areas are given in Figure 9.1.

9.1.1 Urbanized Areas

Urbanized areas are the clearest example of the powerful and usually unbalanced effect of human-induced factors on a geological environment, which very often disturbs hydrogeological safety of the area. Therefore these areas are considered in greater detail.

In the built-up areas, considerable changes in conditions for forming surface- and groundwater discharge occur because the character of their hydraulic connection is disturbed. It should be marked here that if a change of hydrogeological situation in natural conditions is evolutionary by character, it occurs in epochs determined by a geological time scale (millions of years); in urban settings, changes occur by decades (and even by separate years) and hence happen very quickly. This peculiarity is often determined when studying the regularities of groundwater formation in the built-up areas and their interrelations with the environment.

Let's consider separately the factors providing for groundwater replenishment as well as its withdrawal in urbanized areas.

Groundwater replenishment — huge amounts of water, up to 600 l/d per person, are supplied to a built-up area that often considerably exceeds the amount of atmospheric precipitation by 1 m^2.

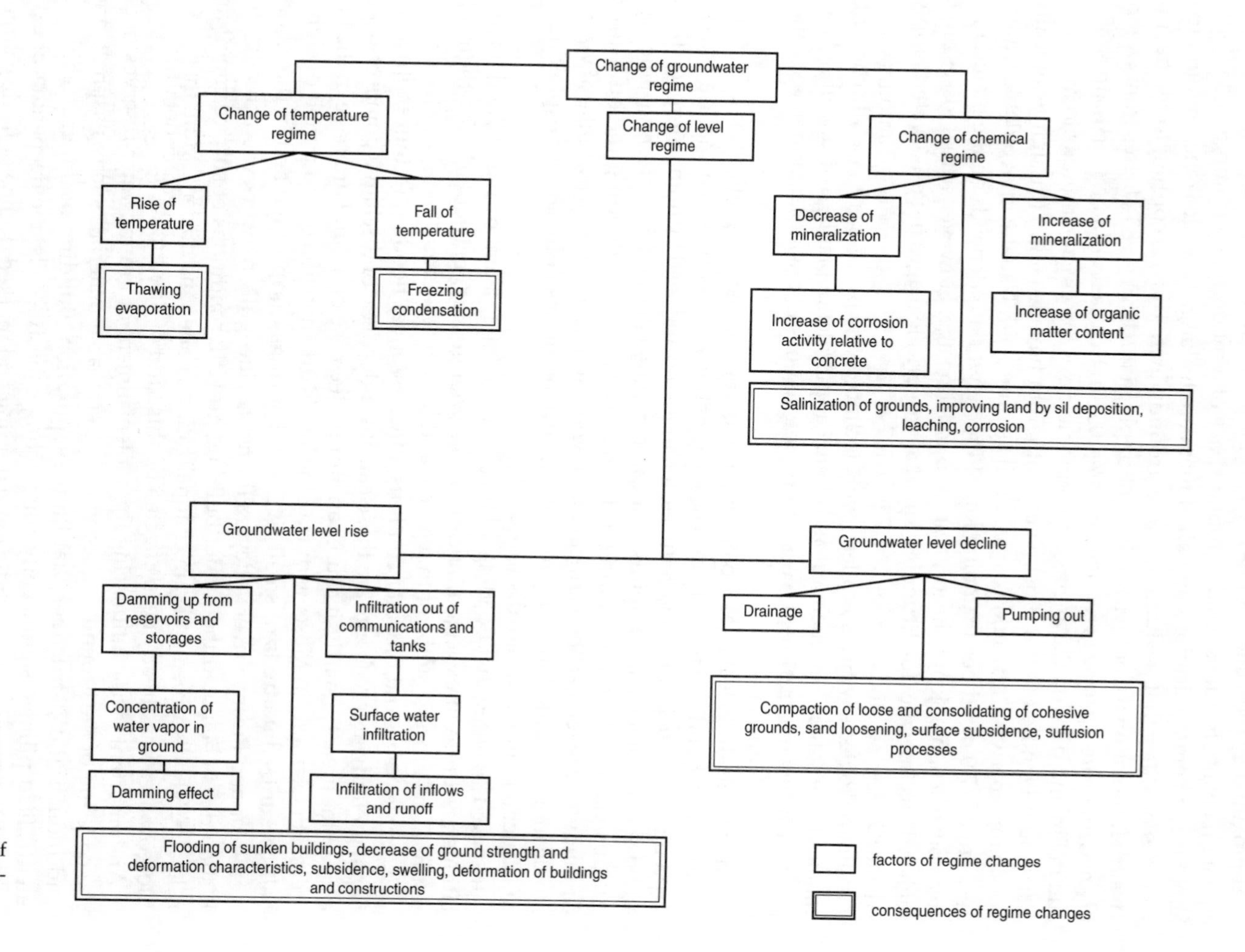

FIGURE 9.1
Scheme of technogenic changes of groundwater regime and their consequences in developed areas.

Approximately the same volume (by 10–15% less) is water drained by domestic and precipitation channelizations. Underground space of a modern city is impregnated with different water communications and all of them are leaking. These leakages amount to huge volumes of water that infiltrate into the groundwater. Thus, according to the data of the Twelfth International Congress on Water Supply in the USA (1983), losses from water-supply networks are up to 30%, for new ones, 7–10%, in Russia for old networks, up to 35–40%, and in Finland, 18% on the average; in separate cases losses can amounts to 34%.

It should be noted that a pipeline cannot fail to have leakages (though technically it is possible); otherwise, it would cost as much as pure gold. Usually unavoidable water leakages are 2.3–7.0 m^3/d per 1 running km of the network. It depends on the material, diameter, construction of joints, head value, and pipeline age. However, actual losses are much higher. Thus, specific (for 1 running km of a network) leakages in Paris' water-supply system are 51.4 m^3/d, and in Tokyo, 57.2 m^3/d.

Industrial objects with a so-called "wet" technological process are as a rule concentrated sources for leakages, i.e., chemical and chemical-petroleum plants, heat and electric-power plants, wastewater-treatment works, pumping stations of a water rotation cycle, and various reservoirs and tanks.

While infiltrating, leakages replenish groundwater, causing significant rise in its level in the form of mounds up to 10–15 m. These formations are occasional by nature and difficult to predict. Here, below leakage sources, technogenic (induced by human impact) lies water and discrete dome-shaped uplifts that can combine later and form technogenic aquifers. All this can result in developing a dangerous process of flooding the built-up areas. Groundwater level reaches a critical point when underground spaces, paths of water communications, foundations, etc. are submerged. Conditions necessary for exploiting certain constructions and areas are disturbed.

It should be noted that flooding when is developing is latent by nature, and its manifestations are quite unexpected, and thus very dangerous, particularly for their adverse results. This process can be both regional and local by nature.

Except leakages out of water conduits, other essential flooding factors and hence additional groundwater recharge are those causing a decrease of discharge points of water balance in the area. Primarily, it causes an abrupt decrease of natural drainage in the built-up areas. For example, in Moscow during the area's urbanization over a period of eight and a half centuries, hundreds of small rivers and springs disappeared and in the Khodynka, Neglinka, and Presnya river valleys, a former ravine–gully network was actually filled up. It should also be noted that under building growth, conditions of surface runoff are as a rule abruptly disturbed and rain channelization is often insufficient. More that 40% of the Moscow area has already been affected by this kind of flooding.

An important factor influencing the replenishment of groundwater reserves in cities is the covering of the ground surface with impermeable materials such as asphalt, and civil and industrial constructions that limit evaporation. Building of impermeable embankments, field pilings, significantly compacting cohesive soils, the deepening of extended constructions results in groundwater damming and marked decline of its discharge.

In general, the process of flooding can be determined as groundwater levels rise toward the land surface or soil moisture increases in the foundations up to and higher than their critical values. It is caused by natural and artificial factors that result in the violation of necessary conditions for constructing and exploiting different objects and the built-up area on the whole, in worsening the ecological situation, and in significant economical losses.

Flooding in cities cannot be only regional and local by character but also it can be an object or point event. In all cases, radical changes of groundwater regime occur in all the flooded sites and groundwater flow is transient. Large industrial complexes, irrigation systems, and water reservoirs can be considered as regional sources within a city. Thus, for instance, in the 1930s, groundwater damming due to the filling of the Khimki water reservoir in Moscow to the height of 20 m, significantly affected the rise of the groundwater level. As a result, some areas in the Moscow river valley were partially submerged. An aquifer in fluvioglacial sands under a moraine changed from an unconfined into a confined aquifer that threatened groundwater by abrupt leakages into trenches and deepened spaces.

Another example of causing a rise in groundwater level at a regional scale can be from fields of piled earth, made during total building up of microrayons. In the pile-field zone, the ground is so compressed that its filtration properties decrease almost by an order. Thus, considerable backing-up due to the damming effect of a pile field occurs and a vast area of flooding appears upstream. Damming of groundwater flow can occur in dug-out foundations of buildings, causing a rise of groundwater level in separate sites.

Regional groundwater level rise (flooding) can be also caused by long interruption of groundwater withdrawal or drainage that results in a level rise of returned groundwater. Thus, for instance, during the Second World War, in the city of Kharkov the well field stopped functioning because of war activities, and this caused the groundwater level to rise by more than 10 m.

If a regional type of flooding (groundwater-level rise) can be taken into account by engineering studies, projecting and consequently exploitation of a built-up area, then a local (object) type of flooding is actually unpredictable. It is mainly connected with errors in building, unsatisfactory exploitation of construction and water communication, and surface runoff disturbance in the area directly adjoining the object. A local type of flooding is widely though discretely spread in cities. Flooding of a separate building occurs quite unexpectedly and it can happen actually in any part of a city. This is due to damage of local water conduits, heat-supplying systems, etc.

Local flooding can be formed by long-standing water in trenches that have filled up with atmospheric precipitation. After building a foundation and filling up a trench, perched water occurs there. As a result, deep spaces are flooded and walls of the ground floor are moistened through seepage.

Moisture accumulation under buildings can occur due to a difference in temperature between the foundation and the surrounding ground. This difference can amount to 5–10°C. With this drop in temperature, water vapor moves out of warm into cold zones and then condenses. This process, in conjunction with a decrease of evaporation in shaded or covered (screened) surfaces can cause local groundwater level to rise, or flooding in corresponding hydrogeological conditions after 7–10 years. This process is insufficiently studied but it is necessary to take it into account when developing a project and exploitation. However, the main reason for flooding is still the leakages from water pipes.

Thus, the process of flooding is caused by multiple factors that form under both natural and artificial factors, including seasonal fluctuations of groundwater level, groundwater cyclicity manifestations, etc.

Areas made up of poorly permeable and inhomogeneous ground, characterized by low drainage are most intensively flooded.

To flood a built-up area, corresponding artificial conditions are needed, that often occur due to the main sources of groundwater recharge. It has been found that the formation of flooding (water logging) occurs at the initial stage of building, while preparing the area for building (vertical planning, road running, disturbing the relief and hence, conditions of surface runoff, digging of trenches and ditches, making different water reservoirs, laying out temporary water communications, etc.). Later in the process of building (laying out of constant underground communications and their testing with simultaneous exploiting of the temporary network, building of foundations, filling up of trenches and ditches with loosened ground, laying out of rain drains, etc.), and consequent exploitation of constructions, a disturbance of existing hydrogeological conditions is intensified.

Generally, flooding of a built-up area has three phases: initial stage (period of building)—local groundwater levels rise under recharge sources in the form of domes (mounds) where chemical composition and temperature changes considerably and technical perched water can be formed. Transient stage (beginning of exploitation)—groundwater domes formed in the initial stage spread and new ones appear under additional sources of recharge. At this stage, groundwater level and soil moisture content can reach critical values. And last, the third stage, when isolated domes combine into a united aquifer. It should be noted that flooding can be stopped by protective measures at any of the singled-out stages.

Based on the above considerations, additional groundwater replenishment in the urbanized areas is, as a rule, accompanied by its increasing pollution. Sources of pollution include networks of industrial, domestic and rain drainage, places of storing industrial and other wastes, and storages of numerous chemical products. Leakages out of these sources together with atmospheric precipitation, saturated with different exhaust and waste gases that, being combined with atmospheric moisture, form acids of corresponding concentrations, for example, sulfuric, nitric, and perchloric, essentially contaminate the groundwater. In this case, soils degrade and vegetation productivity decreases. As a result, groundwater is polluted with heavy metals, petroleum products, and numerous chemical elements. Groundwater pollution is intensified by irrigation systems located close to cities and large industrial facilities.

For urbanized areas, thermal groundwater pollution, caused by heating near-the surface soil layers in places of thermal industrial objects and infiltration of thermal leakages (up to 60–70°C), is also characteristic. In places with a higher temperature, bacteria growth is intensified, which increases oxygen absorption. It should be considered here that surface soil layers in cities are highly saturated by components polluting the groundwater (for instance, toxic microcomponents of arsenic, lead, strontium, cobalt, mercury, cadmium, copper and different radionuclides). That is why even infiltration of treated water leakage out of potable water supply systems often results in groundwater-quality deterioration.

The main peculiarity of forming a groundwater regime in urbanized areas is that it is nonstationary in time and nonhomogeneous in space. It is due to an impact of numerous regime-forming factors in their various combinations, resulting from urban functioning and, primarily, its technical components.

In one and the same built-up area, places with natural (undisturbed or insignificantly disturbed), artificial and mixed regimes can exist. It makes predicting difficult as well as the choice of protective measures. It should be noted that in the same place a level regime can be artificial and temperature and hydrochemical ones — natural and vice versa — groundwater regime in the built-up areas should be considered by separate elements. Depending on the scale of regime-forming factors, its regularities can be local and regional.

It is reasonable to define the concept "groundwater artificial regime." Such a regime or its element occurs when qualitatively new regularities in the groundwater regime occur as a result of dominating artificial factors at a given area and time period. When considering artificial groundwater regime, it is important to assess the natural regime and the possible fluctuations of its parameters. It is reasonable to single out controlled (when using different drainages, well fields etc.) and uncontrolled regimes.

At a single site of an urbanized area, natural and artificial regime regularities can manifest themselves simultaneously (mixed regime) either in a certain sequence, replacing one another, or at different time scales, for instance, natural means annual and artificial means perennial. This groundwater regime should be called a combined one. A cyclic regime is also singled out when artificial and natural regimes consequently replace each other with a certain cyclicity. Inside types and sub-types of an artificial regime, its varieties can be singled out, depending on technical and hydrogeological conditions and the character of transformations that have occurred. For example, take a groundwater regime of squares, yards, different industrial objects, and areas. Only this hierarchic separation of regime makes possible a well-founded prediction with minimum uncertainties, creating a corresponding system of groundwater monitoring and a proven system of engineering protection.

In towns with artificial groundwater level rise, which can cause dangerous engineering-geological and hydrogeological processes, a stressed ecological situation may develop due to:

- Flooding of dug-out spaces, including cellars and subway stations, resulting in moisture and fungal formations on the walls. A favorable environment is formed for mosquitoes and human illnesses, conditions of maintaining and repair of water-, power- and gas-supply systems are markedly worsened and their wear is hastened.
- Flooding of foundation soils, inadmissible decrease of their strength and deformation properties, activization of hazardous geological processes (karst, landslides, fall-downs, suffusion, etc.), that in turn cause dangerous, often catastrophic, deformations of buildings, construction, roads, and engineering systems.
- Increase by 1–2 points of seismicity in the built-up area that results in notable decrease in seismic stability of available constructions and in the necessity of additional and expensive antiseismic measures.
- Increase of groundwater aggressiveness and soil-corrosion activity relative to metal and specific of deepened constructions and communications.
- Inadmissible moistening and salinization of soil in the city, parks, squares, and lawns, causing oppression of vegetation, a rise in price for its maintenance, and often a swamping of the area by polluted drainage wastes.
- Flooding of engineering communication trenches, provoking numerous accidents and causing losses to the environment.
- Pollution of rivers by drainage runoff, flooding of dumps, causing additional polluting of the river runoff and soils.
- Flooding of historical and cultural monuments, destroying of unique historical landscapes, and ancient national memorials.

In the areas where groundwater is polluted with petroleum products, a rise of liquid and gaseous petroleum hydrocarbon toward the land surface initiates dangerously explosive and fire-hazardous situations and worsens sanitary conditions in the area.

On the whole, groundwater level rise deteriorates the sanitary and hygienic state of the urban area, particularly where the water medium provides for the formation of sources of infectious illnesses.

Groundwater withdrawal — in an urban area, due to the necessity of producing the required conditions for building and also exploiting of the urban underground space, creation of an artificial groundwater level decline is made. All this is correlated with the city well fields, which often function in the form of discretely located wells.

Considerably deepened constructions (for instance, subway tunnels), separate buildings, and sites of the area are exploited in obligatory combination with work of constant water-level-declining measures, such as large city drainage systems.

Land-surface screening by buildings, asphalt, and other coverings provides for decreasing groundwater-infiltration recharge. It is known that the density of city buildings accounting for squares and roads pavement, can reach 90% in some city districts. Decrease of groundwater recharge occurs due to removing a part of surface runoff into rain drainage.

Due to intensive groundwater withdrawal, the whole water-balance structure changes in principle and an artificial groundwater regime is formed. However, the natural regime is suppressed by artificial factors. It is preserved in many sites, and in some cases, the natural regime is again formed against a background of artificial ones with regularities and characteristics of a given natural-climatic zone.

It should be marked that groundwater's natural protection considerably decreases in urban areas. This is due to developing the underground space in cities, since poorly permeable layers are cut through and polluted city water can penetrate into an exploited aquifer.

In the cities subjected to intensive groundwater withdrawal, the following unfavorable environmental consequences are observed:

- inflowing of polluted groundwater out of industrial zones, wastewater-treatment sites, and irrigation systems into the well fields
- overdrying and degradation of soils in city parks and squares, loss of moisture-loving plants
- decrease of the river runoff, development of land subsidence
- activization of most geological processes, for instance, karst-suffusion resulting in downfalls and building destruction

Thus, the reaction of an urban area, being a complex natural–technical and social system, is quite different and ambiguous on the groundwater recharge and withdrawal. Groundwater reaction here is not always predictable and is of a probabilistic nature. Groundwater exploitation here is risky, which must be considered when deciding the problem of intensifying groundwater use for different purposes.

Hydrotechnical constructions, and primarily water reservoirs and channels, are powerful regional factors affecting the groundwater and the environment. The zone of their effect amounts to thousands of square kilometers. Due to damming and filtration losses out of reservoirs and channels, the following phenomena are often observed:

- New aquifers are being formed and their regime is caused by conditions of exploiting reservoirs and channels.
- Stable rising of groundwater levels and heads occurs and are distributed over hundreds of kilometers.
- Additional recharging of the main exploited aquifer occurs.
- Flooding (groundwater level rise) and overflooding of the adjacent areas are being developed with all the unfavorable consequences.
- Coastal abrasion, swamping, soil salinization are developed, and, as a result, considerable areas of arable lands, pastures, and different agricultural farms are lost and flora and fauna are oppressed.

9.2 Some Aspects of Assessing Hazard and Risk of Groundwater Use as a Water Supply Source

It is known that sustainable development of the area is primarily connected with ecological–social systems and their natural-resources potential. Groundwater is a very important component of the area's sustainable development. As has already been mentioned, groundwater is more protected from different types of pollution, is less expensive, and readier for use than surface water. That is why groundwater reserves are strategic for human life support. Therefore, they are particularly valuable.

It is very important to find a way of assessing the hazard and risk of groundwater withdrawal in developed areas. It is necessary for providing a maximum and stable water pumping of the layer on one hand and, on the other, for groundwater withdrawal on the environment, including the groundwater itself. There is a certain opposition between the problems to be solved. Really, maximum groundwater withdrawal usually causes activization of negative consequences and vice versa; excluding or weakening these condition will decrease the total yield of the well field.

Thus, a complex and ambiguous problem of optimizing a well-field exploitation appears, aimed at making acceptable a level of possible risk. Assessing groundwater resources under increasing anthropogenic activities becomes particularly urgent now and requires special investigations. The results of this work make it possible to prove necessary preventive measures for controlling risk on the basis of assessing probable losses (ecological, social, economic) in advance.

Let's consider some aspects of assessing hazard and risk under groundwater well-field exploitation.

Hazard means an available or possible (potential) threat of damaging a certain object. Hazard (H) is characterized by different direct or indirect attributes (I = index of hazard). These effects have their own sources with intensity W and frequency or probability of originating P. Hazardous effects develop according to some scenarios Sc., i.e., and there is a function Im (Sc,W,P). Hazard should be considered as a certain warning (prevention) against a possible threat or unfavorable results of decisions made by a responsible person.

An object of hazard should then be identified. This is the object that is finally threatened or can be threatened by the actions of a decision-maker. In this case, two objects of hazard are singled out — groundwater resources and environment. It should be marked that a hazard and an object of hazard are an inseparable couple. The reason is man, and, more exactly, his activities. Hazard is created (initiated) by man, realizing his decision.

It is necessary to support a safe engineering–geological situation in the developed area. Safety is achieved by predicting, preventing, decreasing, and timely interruption of dangerous effects, increasing safety (in this case protection) of the hazard object itself, and use of its protection in the form of adaptational and controlled regulation, and also by decreasing risk in human activities to a permissible limit. A subject person being active or passive initiates hazardous conditions. There appears a risk of damaging the environment, an object and himself. Therefore, risk is a measure of possibility (probability) for realizing hazard in the form of some damage (social, ecological, physical) while making decisions by a person, i.e., the situation artificially created by that person (Dzektser, E. S., 1992).

To provide safety after strict determination of the objects of hazard, it is necessary to single out sources of hazard and risk (available and potential), or bearers of hazard and risk. Generally, environment, object, and subject are considered to be such bearers (Dzektser, E. S., 1994).

It should be noted that the problem formulated here requires solving of low, structured problems, i.e., problems having quantitative and qualitative variables. Accepting of adequate solutions is very difficult for the person under such conditions, since a series of uncertainties should be overcome.

Let's consider a sequence of risk formation. This will allow a proper moment for applying necessary management measures to be taken for preventing and protecting an object from hazardous impact, for possible risk reduction by making a safe engineering–hydrogeological situation well-field decision.

The problem considered is subdivided into two interconnected ones: the first is the reliable withdrawal of water needed from the layer. The second is the reliable provision of necessary water inflow into the well field, i.e., a steady supply that solves the first problem a person should initiate in a well field, i.e., a steady supply with a given water amount.

When beginning to solve the first problem, one should initiate a well-field construction or necessary yield changes (g) under reconstructing an available one. Thus, a possible scenario of these activities Sc, i.e., $I_m = S_c$ is accepted as a characteristic of the hazard index.

The probability of manifesting the q event is possible if there are some activities developing by one of the possible scenarios Sci with probability P (S_{ci}) in the input of the system. Then, basing the formula for a full probability, the following equation for assessing risk of possible well-field impact on the groundwater is

$$\mathrm{R(q)} = \mathrm{P(q)} = \sum\nolimits_{i=1}^{n} P(q/S_{ci})P(S_{ci}) \tag{9.1}$$

where P (q/S_{ci}) — is an assumed probability of event q under condition, that the scenario of the human activities in connection with a well field will be S_{ci}.

Often, constructing a new or reconstructing an available well field for a certain period is a problem solved in advance. Then P (S_{ci}) ≈ 1 and R (q) = P (q/S_{ci}), i.e., P (I_m) = R (q). In this case, the probabilistic character of a well-field's impact on the groundwater will be preserved and will depend on the quality of the field project, on possible failures in the pumping wells (pump, filter, etc.), and on damages in water and power supply lines.

The meaning P (q) is often accepted on the basis of available experience in well fields. Reserved water pumping wells are often used as insurance. However, there is no absolute safety and a certain risk in the well-field operation always remains. It mostly depends on the quality of managing the water-pumping regime, determined by a person. One successful way to lower risk is through a complex use of surface and groundwater.

When water is pumped out of a layer, groundwater level (head) S decreases. Here S = f (q, d, t, r) where a is filtration parameters (layer transmissivity, piezoconductivity, filtration coefficient, water yield, etc.), t is duration of well-field functioning, r is a coordinate of the point in the layer in axisymmetric problem. Here the necessity of considering the second problem appears.

A possible level (head) decline to a given value in the layer where there is a well field can cause some unfavorable consequences at considerable distances for both the groundwater and the environment. Therefore, a proven definition of permissible groundwater-level decline values (S_{cr}) that is a threshold of hydrogeological safety for a given water-bearing layer and environment.

It is known that a level (head) decline in the layer can cause polluted water inleakage to a well field, a decrease in groundwater protection, and changes in the values of aquifer recharge and discharge.

Groundwater-level decline can cause lowering of productivity, overdrawing of agricultural lands, forests, swamps, and desertification in arid zones.

Due to the above-mentioned uncertainties, all predictive calculations have elements of risk. The problem is in lowering the risk without liquidating it.

Let's consider a principal approach to solving a multifactor, ambiguous, and often conflicting situation that appears in the process of well-field functioning between acting well-field requirements and environment protection.

The first problem to be solved is the choosing of an optimal regime for a well-field function. For instance, there is a danger of polluting the groundwater. In case of a pollutant's escaping from the layer into a system of potable-water supply, there appears a certain social problem in the form of people poisoning. It is necessary to decrease the risk of possible poisoning to an acceptable level by special measures for managing the risk. This will require certain capital investments.

Decrease of pollutant inleakage by lessening q and, hence, S to $q = f(S_{cr})$ or building up of adequate purifying constructions can be such measures. Decrease of the well-field yield by D q will require additional expenditures for using some other water resources, for instance, from a neighboring region.

So the problem is to determine the volume of such expenditures for the measures to be taken up that do not exceed possible losses from pollution within a permissible risk framework.

If the situation is not managed, the risk of polluting the layer and the pollutant's getting into a water supply system R (C) is determined in the following way:

$$R(C) = \sum_{i=1}^{n} P(C/Si)P(Si)L \tag{9.2}$$

P (C/Si) is an assumed probability of pollutant with concentration C getting into the layer, under its level (head) decline. In this case, $C > C_o$ will occur (C_o is maximum permissible concentration); L is a social loss in economic equivalent.

Equation 9.2 actually determines a level of the layer vulnerability. P (S_i) is a probability of Si decrease in a given point of the layer. Risk in the form of probabilistic expenditures for measures, decreasing pollutant concentration in the well field R_p (C) by decreasing S by Δ S, is also determined by Equation 9.2, but instead of S a value $S_p = S - \Delta S$ is used.

Joint curved plotting for R and Rp by function Equation 9.2 will allow one to determine the value of S when expenditures for measures will be less or equal to the permissible risk. The task is solved by a method of successive approximations.

Then problems on interaction between a well field and the environment can be solved. The area within a functioning well-field zone is subdivided by S_{cr}, considering environmental requirements. In places where the level of risk in the well field, as projected, exceeds a permissible one for the environment, compromise solutions should be assumed, up to solution of it.

Conclusion

In conclusion, it should be once again noted that the problem of groundwater use is a composite part of a common problem of rational natural use and environmental protection. Only a joint consideration of all the aspects of interaction between the groundwater and other environmental components can make it possible to elaborate on a long-term program for rational groundwater use and protection.

Natural protection restrictions for groundwater withdrawal must be considered and possible changes in groundwater resources under the impact of engineering and economic activities must be investigated. It is particularly important to work out predictions of increased groundwater pumping for centralized water supply of a population and for industry and agriculture. It should again be stressed that the task of specialists at present is not only to calculate the water volume that can be pumped out of an aquifer in specific hydrogeological conditions during a certain time period, but also to assess the possible changes in different environmental components that may be caused by the withdrawal of groundwater. As a result, specialists must prove and recommend, if necessary, special measures to minimize possible negative consequences of groundwater withdrawal, particularly when exploiting large well fields.

Determining the function of groundwater in total water resources and water balance of separate regions — river basins, lakes, and seas — is a separate, but no less important, problem.

Solving the problems mentioned will absolutely provide for increasing effectiveness and rationality of groundwater use and make it possible for decision makers to prove modern and prospective projects for the water supply of separate regions.

In this book, I have made only a first attempt to consider and generalize the experience available in different countries for studying the interaction between groundwater and the environment. I know that a number of aspects are only briefly considered in the book — in some cases, only in general. This is for two main reasons, the first being the limited size of the book and also the great number of detailed publications for separate, specific areas (this I tried to compensate for by providing an extensive list of references). The second reason is the shortage of scientific and methodical development of a number of the aspects. Thus, in conclusion, it is reasonable to briefly formulate the main tasks of further scientific and practical investigations on the problem considered.

These tasks are the following:

- to improve the available and to develop new methods for assessing groundwater resources accounting for natural measures
- to develop and put into practice nature-protecting criteria determining the acceptable impact of groundwater withdrawal on other components of the environment, and also the acceptable effect of anthropogenic activities on groundwater resources and quality
- to perfect the available and to develop new methods for predicting changes in groundwater resources and quality under intensive anthropogenic activities and possible climate changes

- to substantiate the principles of conducting groundwater monitoring in different natural-climatic and anthropogenic conditions as a component of the general monitoring of water resources and the environment
- to improve methods of assessing groundwater vulnerability to pollution in the main aquifers used for water supply
- to perfect methods of artificial groundwater recharge and to use them more widely in active well fields
- to develop mathematical models of interaction between ground- and marine water in different geologic-hydrogeologic conditions of the coastal zones and also methods for predicting marine-water intrusion into the aquifers under intensified groundwater withdrawal by coastal well fields
- to perfect the available and develop new methods for quantitative assessing groundwater impact on the hydrochemical, hydrobiological, and thermal regimes of lacustrine and marine water
- to assess the function of groundwater discharge in the water–salt balance of separate seas and large lakes

References

Allen, R.H., Gottlieb, M., Clute, E., Pongsiri, M.J., Sherman, J., and Obrams, G.I., Breast cancer and pesticides in Hawaii: the need for further study. *Environ. Health. Perspect.*, 1997 Apr; 105 Suppl 3: 679-83.

Barends, Frans B.J., 1995. Land subsidence by fluid withdrawal by solid extraction — theory and modeling; environmental effect and remedial measures. *IAHS Publ.*, N 234, pp. VII-XIV.

Batoyan, V.V. and Brusilovsky, S.A., 1976. Fresh water on the bottom of the Black Sea. Priroda #2, p. 17-22.

Be'er, S.A., Parasitologic Monitoring in Russia (concept framework). *Med. Parasitol.* 1996. 1. p. 3-8.

Bodelle, J. and Margat, J., 1980. *Groundwater in France.* Paris, Masson. 208 p.

Borevsky, B.V., Grodzensky, V.D., and Yazvin, L.S., 1989. Assessing groundwater reserves. Kiev, Vyshaya Shkola, 407 p.

Borevsky, B.V., Grodzenskii, V.D., Yazvin, L.S., and Zektser, I.S., 1987. Formation study and use of groundwater resources for water supply of towns and urban agglomerations: urgent problems. Moscow, Nauka, p. 11-14.

Borevsky, B.V., Zeegofer, Yu. O., and Zektser, I.S., 1989. Problems of water supply in Moscow and Moscow region. Moscow, *Nauka,* p. 111.

Borevsky, B.V., Zeegofer, Yu.O., Pashkovsky, I.S., and Yazvin, L.S., 1989. Groundwater resources and quality of Moscow region. *Nauka.* p. 111-120.

Borevsky, B.V. and Yazvin, L.S., 1991. Ecological aspects of exploring and assessing groundwater safe yield for potable and industrial needs. *Hydrogeological Aspects of Ecology.* Moscow, VSEGINGEO. p. 40.

Borevsky, L.V., Borevsky, M.V., and Mironenko, V.A., 1994. Strategy of water objects protection from groundwater pollution with petroleum products. *Proc. of Congress on Water: Ecology and Technology,* Moscow. p. 243-245.

Bro-Rasmussen, F., 1996. Contamination by persistent chemicals in food chain and human health. *Sci. Total Environ.* Sept; 188 Suppl 1: S 45-60.

Bruner, M.A., Rao, M., Dumont, J.N., Hull, M., Jones, T., and Bantle, J.A., Ground and surface water developmental toxicity at a municipal landfill: discription and weather-related variation. *Ecotoxicol. Environ. Saf.* 1998 Mar; 39(3): 215-26.

Brusilovsky, S.A., 1971. On the possibility of estimating submarine discharge by its geochemical manifestation. Comples investigations of the Caspian Sea. Moscow, #2, MGU, p. 68-74.

Brusilovsky, S.A. and Glazovsky, N.F. 1983. Problems of the hydrogeology of the ocean. Hydrodynamics and Sedimentation. Moscow, Nedra. p. 66-73.

Cable, J., Bugne, C., Burnett, W., and Chanton, J., 1996. Application of ^{222}Radon and CH4 for assessment of groundwater discharge to the coastal ocean. *Limnol. Oceanogr.* #41 (6), p. 1347-1353.

Cantor, K.R., Drinking water and cancer. *Cancer Causes Control* 1997 May; 8 (3): 292-308.

Cherepansky, M.M., Kulbeda, I.P., and Usenko, V.S., 1987. Assessing the impact of groundwater on the river runoff considering evaporation. Moscow. NII Hydrotechnique and Amelioration. 138 p.

Chian-Min Wu, 1992. Groundwater development and management in Taiwan. J. *Geol. Soc. China.* vol. 35, N 3, pp. 293-311.

Chow, V.T. (Ed.), 1964. *Handbook of Applied Hydrology.* McGraw-Hill, New York.

Custodio, E., 1982. Elements of groundwater flow balance (natural and as affected by man). Intern Symp. Computation of Groundwater Balance, Varna, 17 p.

Dogdeim, S.M., Mohamed, el-Z., Gad-Alla, S.A., el-Saied, S., Emel, S.Y., Mohsen, A.M., and Fahmy, S.M., Monitoring of pesticide residues in human milk, soil, water, and food samples collected from Kafr El-Zayat Governorate. *J AOAC Int.* 1996 Jan-Feb; 79(1): 111-6.

Dormrachev, G.I. and Grishina, I.N., 1987. Current problems in engineering geology and hydrogeology of cities and urban agglomerations. Moscow, Nedra, p. 209.

Dourison, M.C. and é Felter, S.P., Route-to-route extrapolation of the toxic potency of MTBE. *Risk Anal.* 1997, Dec; 17(6): 717-25.

Durazo, J. Farvolden, R.N., 1989. The groundwater regime of the valley of Mexico from historic evidence and field observations. *J. Hydrol.* vol. 112, N 1-2, p. 171-190.

Dzektser, E.S., 1992. Geological hazard and risk. *Eng. Geol.* #6. p. 3-10.

Dzektser, E.S. 1994. Methodological aspects of geological hazard and risk. *Geoecology,* #3. p. 3-10.

Dzhamalov, R.G., 1973. Groundwater flow of the Terek-Kuma artesian basin. Moscow, Nauka, 142 p.

Dzhamalov, R.G., Zektser, I.S., and Meskheteli, A.V., 1977. Groundwater discharge into seas and world oceans. Moscow, Nauka, 94 p.

Dzhamalov, R.G., Zektser, I.S., and Ivanov, V.A., 1978. Studying groundwater from cosmos. *The Earth and Universe,* #2. p. 11-17.

Dzhamalov, R.G., Zektser, I.S., and Zlobina, V.L. 1996. Groundwater contaminationa and acidification in Russia. Symposium on Water and Global Pollution, Seoul, South Korea. p. 12-19.

Economic Bull. 1982. Water Economy Prospects for 1990 and 2000., vol. 34, No. 1, p. 91-117.

Eddleston M., 1996. Structural damage associated with land subsidence caused by deep well pumping in Bangkok, Thailand. *Qwar. J. Eng. Geol.*, vol. 29, N 1, pp. 1-4.

Elpiner, L.I., On the Influence of the Water Factor on the Health of Russia's Population. *Water Res.*, 1995, v. 22, N 4, p. 418.

Elpiner, L.I. and Delitsin, V.M., Medico-biological problems of the Aral catastrophy. *Proceedings of the USSR Academy of Sciences.* #4, July-Aug. 1991. 103-12.

Elpiner L.I., Shapovalov, A.E., and Zeegofer, Y.O., Subterranean water under conditions of intensive technogenes: hydroecological and medical aspects *Mrlioratsya i vodnoe khozaistvo* 1998, #3, May-Jun, p. 66-67.

Elpiner, L.I, and Vasiliev V.S., *Problems of Drinking Water Supply* in USA. Nauka, 1983, 168 p.

Environmental Quality. 1978. The Ninth Annual. Rep. Of the Counc. On Environ. Quality. Washington, GPO, 1979, xii+599 p.

Everett, L.G., Wilson, L.G., and Hoylman, E.W., 1984. Vadose zone monitoring for hazardous waste sites. Noyes Data Corp. USA, 358 p.

Foster A., 1995. Active ground fissures in Xian, China. *Quart. J. Eng. Geol.*, vol. 28, N 1, pp. 1-4.

Freeze, R.A. and Cherry, J.A., 1979. *Groundwater.* Prentice-Hall Inc., Englewood Cliffs, 604 p.

Geographic information system database for geohydrologic assessments, south San Francisco Bay and Peninsula area, California, vol. 79, N 4, p. 584.

Glazovsky, N.F., Ivanov, V.A., and Meskheteli, A.V., 1973. On studying submarine springs. Okeanologia, Vol. 13, #2. p. 249-254.

Goimerac, T., Kartal, B., Bilandzic, N., Roic, B., and Rajcovic-Janje, R., Seasonal atrazine contamination of drinking water in pig-breeding farm surroundings in agricultural and industrial areas of Croatia. *Bull. Environ. Contam. Toxicol.* 1996 Feb; 56 (2) : 225-30.

Gottlieb, M.S., Carr, J.K., and Morris, D.T., Cancer and drinking water in Louisiana: colon and rectum. *Int. J. Epidemiol.* 1981 Jun; 10 (2); 117-25.

Greenfield, B.J., 1978. Groundwater storage in unconfined aquifers. The Thames conservancy division model. *Publ. Thames Water,* 15 p.

Groundwater discharge in the USSR territory. 1966. Moscow, MGU. 303 p.

Groundwater Report, 1996. HGCSD.

Guidelines for drinking-water quality. Second ed. V.1; Geneva, WHO, 1993.

Hamberg, D., 1989. The role of groundwater in Israel's integrated water system. *IAHS Publ.*, N 188, pp. 501-514.

Haryomo, 1995. Relation between groundwater withdrawal and land subsidence in Kalantan, Malaysia. *IAHS Publ.*, N 234, pp. 31-33.

Haupert, T.A., Wiersma, J.H., and Goldring, J.M., Health effects of ingesting arsenic-contaminated groundwater. *Wis. Med. J.* 1996 Feb; 95 (2): 100-4.

Hildesheim, M.E., Cantor, K.P., Lynch, C.F., Dosemeci, M., Lubin, J., Alavanja M., and Craun, G., Drinking water source and chlorination byproducts. II. Risk of colon and rectal cancers. *Epidemiology* 1998 Jan; 9(1): 29-35

Hiller J.R., 1993. Conf. "Aqulfera at Risk Towards Nat. Groundwater Qual. Perapest." Canberra 13-15 February, 1993. AGSO J. *Austral. Geol. and Geophys.*, vol. 14, N 2-3, pp. 213-217.

Hopenhayn-Rich, C., Biggs, M.L., Fuchs, A., Bergoglio, R., Tello, E.E., Nicolli, H., and Smith, A.H., Bladder cancer mortality associated with arsenic in drinking water in Argentina. *Epidemiology.* 1996. Mar; 7 (2): 117-24.

Hydrogeology of the USSR. Collected volume. Issue 3. Groundwater resources of the USSR and perspectives for their use. 1973. Moscow, Nedra, 279 p.

Ijsselmuiden, C.B., Gaydos, C., Feighner, B., Novakosky, W.L., Serwadda, D., Caris, L.H., Vlahov, D., and Comstock, G.W., Cancer of the pancreas and drinking water: a population-based case-control study in Washington County, Maryland. *Am. J. Epidemiol.* 1992, Oct 1; 136 (7): 836-42.

Kats, D.M. and Pashkovskii, I.S., 1988. *Meliorative Hydrogeology.* Moscow. Agropromizdat.

Khublaryan, M.G., Putyrskii, V.E., and Frolov, A.P., 1987. Modeling of conservancy divisional model flow in the unsaturated-saturated soil-river system. *Proc. International Symposium on Groundwater Monitoring and Management*, Dresden.

Kochetkov, M.V. and Yazvin, L.S., 1992. Function of groundwater in potable water supply. Mater. Seminar "Improving technological and sanitary reliability of systems for potable and domestic water supply." Moscow.

Kochetkov, M.V. and Yazvin, L.S., 1992. *Proc. Workshop on enhancing the technological and sanitary safety of public water-supply systems.* Moscow, Dom Nauchno-tekhn. Propagandy, p. 43.

Kovalevsky, V.S., 1993. Formation and long-term variations in the water regime on the East European plain. Moscow, *Nauka*, p. 110.

Kovalevsky, V.S., 1994. Effect of changes in hydrogeological conditions on the environment. Moscow, *Nauka.*

Krainov, S.R. and Shvets, V.M., 1987. Geochemistry of groundwater for potable and domestic purposes. Moscow, Nedra, p. 237.

Krainov, S.R. and Shvets, V.M., Geochemistry of groundwater used for household and drinking water supply, 1997, Moscow: Nedra, 237 pp.)

Krasovsky, G.N. and Yegorova, N.A., Principles and Criteria of a New Concept of Water Quality Control. In: *Environmental Hygiene.* Moscow, 1990, p. 141.

Kroop, R.H. and Nocido, M.E., 1988. Water Supply Critical areas. Water World Dev.: Proc. 6th *IWRA World Congr. Water Resour.*, Ottawa, May 22-June 3, 1988, vol. 2, Urbana, pp. 34-44.

Kuchment, L.S., Demidov, V.N., and Motovilov, Yu.G., 1983. Formation of river runoff. *Nauka,* Moscow, p. 215.

Kudelin, B.I., 1960. Principles of regional assessment of natural groundwater resources. MGU Publ., Moscow, p. 344.

Kukkula, M., Arstila, P., Klossner, M.L., Maunula, L., Bonsdorff, C.H., and Jaatinen, P., Waterborne outbreak of viral gastroenteritis. *Scand. J. Infect. Dis.* 1997; 29(4): 415-8.

Kulakov, V.V., 1990. Fresh groundwater reservoirs in the Amur Area. Vladivostok: Dal'nevostochn. Otd. Akad. Nauk SSSR.

Kuzma, R.J., Kuzma, C.M., and Buncher, C.R., Ohio drinking water source and cancer rates *Am. J. Pub. Health.* 1977. Aug; 725-9

Lacey, R.F. and Shaper, A.G., Changes in water hardness and cardiovascular death rates. *Int. J. Epidemiol.* 1984. Mar; 13 (1):18-24

Lebedeva, N.A., 1972. Natural groundwater resources of the Moscow artesian basin. Nedra, Moscow, p. 148.

Linsley, R.K., Konler, M.A., and Paulhus, J.L.H., 1962. Applied hydrology. Leningrad, Gidrometeoizdat, p. 756.

Lukjanchikov, V.M., Pollution of Groundwater with Oil: Scale, Detection Procedure, Monitoring, Rationality of Protective Measures. *Proc. Int. Congr. "Water: Ecology and Technology." Moscow, "Sibico International,"* 1996, p. 140.

Manual for Drinking Water Quality Control. 2nd Edition, v. 1, WHO, Geneva, 1994, p. 250.

Manual for Forecasting Medicobiological Consequences of Hydraulic Works Construction. L.I. Elpiner and S.A., Bear, Eds.

Manual for hygiene of water supply. S.N.Cherkinsky, Ed., Meditsina, 1975. 328 p.

Margat, J., 1982. Development or overdevelopment of groundwater reserves? International Symposium on the Computation of Groundwater Balance, Varna, Bulgaria, p. 10.
Masironi, R., Pisa, Z., and Clayton, D., Myocardial infarction and water hardness in European towns. *J. Environ. Pathol. Toxicol.* 1980. Sep; 4 (2-3): 77-87.
McLachlan, D.R., Bergeron, C., Smith, J.E., Boomer, D., and Rifat, S.L., Risk for neuropathologically confirmed Alzheimer`s disease and residual aluminum in municipal drinking water employing weighted residental histories. *Neurology.* 1996. Feb; 46 (2): 401-5.
Melloul, A. and Collin, M., 1994. *Inst. J. Earth Sei.*, vol. 43, N 2, pp. 105-116.
Methodological recommendations for assessment of the safe yield of drainage groundwater in solid mineral deposits, 1992. Moscow, *VSEGINGEO*, 45p.
Mirzaev, S.Sh. and Bakusheva, L.P., 1979. Assessment of the effect of water management measures on groundwater resources. Tashkent, Fan.
Mishra, S.K. and Singh, R.P., 1993. Prediction of subsidence in the Indo-Gangetic basin carried by groundwater withdrawal. *Eng. Geol.*, vol. 33, N 3, pp 227-239.
Moore, A.C., Herwaldt, B.L., Craun, G.F., Calderon, R.L., Highsmith, A.K., and Juranek, D.D., Surveillance for waterborne disease outbreaks – United States, 1991-1992 MMWR CDC Surveill. Summ. 1993. Nov 19; 42(5): 1-22.
Morin, M.M., Sharrett, A.R., Bailey, K.R., and Fabsits, R.R., Drinking water source and mortality in U.S. cities. *Int. J. Epidemiol.* 1985. Jun; 14 (2): 254-64.
Najem, G.R., Strunck, T., and Feuerman, M., Health effects of a Superfund hazardous chemical waste disposal site. *Am. J. Prev. Med.* 1994. May-Jun; 10 (3): 151-5.
National Primary Drinking Water Regulation. Fed. Register, 1991, 30 Jan.
Oborin, A.A., Mikhailov, G.K., Karebanova, I.G., and Rubinshtein, L.M., 1994. *Proc. Int. Congress on Water Ecology and Technology.* Moscow, vol. 1, p. 259.
Ogilvy, N.A. and Semendyaeva, L.V., 1972. A hydrodynamic model of a system of artesian aquifers using geophysical information. In: *Groundwater Flow and its Study.* Moscow, p. 88-100.
On the State of Water Supply to the Population in Russia and Measures to Improve the Quality of Drinking Water. *Proc. of Interdepartmental Commission on Ecological Safety* (Sept. 1994 - Oct. 1995) Russia's Ecological Safety, Issue 2, Section 9, Moscow, 1996, Yuridichaskaya Literatura, p. 178-190.
Pinigin M.A., Cherepov, Y.M., and Elpiner, L.I. Proc. *Third Int. Congr. "Water: Ecology and Technology."* Moscow, 1998. p. 632-633.
Plitman, S.I., Evaluation of Hygienic Efficiency of Water Protection Measures. Methodological Recommendations. Approved by RSFSR Ministry of Health, 1988, Moscow, 1989, 18 p.
Plotnikov, N.I., 1989. Human-induced changes in hydrogeological conditions. Moscow, Nedra.
Potie, L., 1973. Investigations and capture of submarine freshwater springs. Submarine springs at Port Miou, Cassis, France. Second Int. Symp. Groundwater Palermo, 30 p.
Pozin S.G., On the hygienic significance of particular factors affecting water quality in communal and drinking water supply. *Proc. Int. Congr. "Water: Ecology and Technology."* Moscow, 1998. p. 633-634.
Prokhorov, B.B. and Revitch, B.A., Medicodemographic Situation in Russia and the State of the Environment. Working Reports of Russian Academy of Sciences, Moscow, Center for Demography and Human Ecology, 1992, Issue 6, 26 p.
Punsar, S. and Karvonen, M.J., Drinking water quality and sudden death: observations from West and East Finland. *Cardiology.* 1979. 64 (1), 24-34.
Rakhmanin, Yu.A., Mikhailova, R.I., Monisov, A.A., Rogovets, A.I., and Cheskis, A.B., Regional Peculiarities of Drinking Water Quality in Russia and Current Procedure of Its Integrated Hygienic Evaluation. In: *Regional Problems of Health Management in Russia,* V.D. Belyakov, Ed., Moscow, Russian Ascademy of Sciences, 1996, p. 162-171.
Revis, N.W., Major, T.C., and Norton, C.Y., The effect of calcium, magnesium, lead, or cadmium on lipoprotein metabolism and atherosclerosis. *J. Environ. Pathol. and Toxicol.*, 1980, Sept., vol. 4, # 2/3, p. 293-304.
Richter, E.D. and Safi, J., Pesticide use, exposure, and risk: A joint Israeli-Palestinian perspective. *Environ Res,* 1997; 73 (1-2): 211-8.

Rose, J.B. Environmental ecology of Criptosporidium and public health implications. *Annu, Rev. Public Health.* 1997; 18: 135-61.

Rubenowitz, E., Axelsson, G., and Rylander, R., Magnesium in drinking water and death from acute myocardial infarction. *Am. J. Epidemiol.* 1996. Mar 1; 143 (5): 456-62.

Sakalauskene, D.J. and Célene, R.E., 1977. Problems of generation of groundwater resources in the Southern Baltics. Vilnius, Mintis.

Sakamoto, N., Shimizu, M., Wakabayashi, I., and Sakamoto, K., Relationship between mortality rate of stomach cancer and cerebrovascular disease and concentrations of magnesium and calcium in well water in Hyogo prefecture. *Magnes. Res.* 1997. Sep; 10 (3): 215-23.

Sanitary Rules and Standards. SanPiN 2.1.4.559-96. Drinking Water. Hygienic Requirements for Water Quality in Centralized Systems of Water Supply. Quality Control. Moscow, 1996, Goskomepidnadzor Rossii. 111 p.

Semenov, S.V., Monisov A.A., Rogovets A.I., and Elpiner L.I., 1994. Hygienic Problems of Water Supply to the Population. Melioracia I vodnoe khozaistvo, 1994, N 5, p. 40-42.

Shevelev, F.A. and Orlov, G.A., 1987. Water supply of large towns abroad. Moscow, *Stroiizdat*, 351 pp.

Sonneborn, M. and Mandelkow, J., German studies on health effects of inorganic drinking water constituents, *Sci. Total Environ.* 1981. Apr, 18:47-60.

Spink A.E.F. and Welson E.E.M., 1990. Groundwater resource management in coastal aquifers. IAHS Publ., N 173, pp. 475-484.

Stacpoole, P.W., Henderson, G.N., Yan, Z., and James, M.O., Clinical pharmacology and toxicology of Dichloroacetate. *Environ. Health Perspect.* 1998 Aug; 106 Suppl. 4:989-94

Stern, B.R. and Tardiff, R.G., Risk characterization of methyl tertiary butyl ether (MTBE) in tap water. *Risk Anal.* 1997. Dec; 17 (6): 727-43.

Testa, S.M., 1991. Elevation changes associated with groundwater withdrawal and reinjection in the Wilmington area, Los Angeles coastal plain, California. IAHS Publ., N 200, pp. 485-502.

Tohyama, E., Relationship between fluoride concentration in drinking water and mortality rate from uterine cancer in Okinawa prefecture, Japan *J. Epidemiol.* 1996. Dec; 6(4): 184-91.

Vsevolozhsky, V.A. and Dyunin, V.I., 1973. On the direction of pore water migration in compacting sediments. In: Interaction of Surface and Groundwater Runoff. MGU Publ., Moscow, p. 186-198.

Water Economy prospects for 1990 and 2000. *Econ. Bull. for Eur.* 1982. vol. 34, No. 1, p. 91-117.

Yamamoto, S., 1986. Groundwater resources in Japan with special reference to its use and conservation. IAHS Publ., N 151, pp. 381-389.

Yang C. and Hung C., Colon cancer mortality and total hardness levels in Taiwan`s drinking water. *Arch. Environ. Contam. Toxicol.* 1998. Jul; 35 (1):148-51.

Yang, C.Y., Cheng, M.F., Tsai, S.S., and Hsieh, Y.L., Calcium, magnesium, and nitrate in drinking water and gastric cancer mortality. Japan *J. Cancer Res.* 1998. Feb; 89(2): 124-30.

Yang, C.Y., Chiu, J.F, Chiu, H.F., Wang, T.N., Lee, C.H., and Ko, Y.C. Relationship between water hardness and coronary mortality in Taiwan. *J. Toxicol. Environ. Health.* 1996. Sep; 49 (1); 1-9

Yang, R.S., el-Masri, H.A., Thomas, R.S., Constan, A.A., and Tessari, J.D., The application of physiologically based pharmacokinetic/pharmacodynamic (PBPK/PD) modeling for exploring risk assessment approaches of chemical mixtures. *Toxicol. Lett.* 1995. Sep; 79(1-3): 193-200.

Yazvin, L.S., Borevsky, B.V., Kochetkov, M.V., and Zektser, I.S., 1994. Groundwater use intensification as a method for enhancing the dependability of public water supply systems. International Congress "Water: Ecology and Technology", Moscow, p. 565-573.

Yazvin, L.S. and Zektser, I.S., 1996a. Russia's fresh groundwater resources: present state, development prospects and investigation goals. *Water Resources*, vol. 23, No. 1, p. 24-30.

Yazvin, L.S. and Zetsker, I.S., 1996b. Variations in groundwater resources under human impact. *Water Resources*. #2, p. 75-81.

Yokoyama, T. and Itadera, K., 1995. Cause of local subsidence in the Sagamigawa alluvial plain, Kanagawa, Japan. IAHS Publ., N 234, pp. 79-87.

Zeegofer, Yu O., Vilkovich, R.V., and Daniluchev, N.V., 1998, Federal ecological program, Volga River in Moscow region. Moscow, 201 p.

Zektser, I.S., 1968. Fresh groundwater natural resources in the Baltic countries. Moscow, Nedra, 105 p.

Zektser, I.S., 1977. Laws of formation of groundwater principles of its study. Moscow, Nedra Publ., 1973 p.

Zektser, I.S., Plotnikov, N.I., and Yazvin, L.S., 1979. On perspectives of groundwater use. *Water Resour.* #2, p. 75.

Zektser, I.S., Dzhamalov, R.G., and Meskheteli, A.V., 1984. Subsurface water exchange between land and sea. Leningrad. *Gidrometeoizdat*, 207 p.

Zelenin, I.V., 1972. Natural groundwater resources of Moldavia, Kishinev, Shtiintsa, 214. p.

Zemla, B., Geography of the incidence of stomach cancer in relation to hardness of drinking water and water supply. *Wiad Lek.* 1980. Jul 1. 33 (13):1027-31.

Zhorov, A.A., 1995. Groundwater and environment. Moscow. 136 p.

Zielhuis, R.L. and Haring, B.J., Water hardness and mortality in the Netherlands. *Sci Total Environ* 1981. Apr; 18:35-45.

Zlobina, V.L., 1985. The influence of groundwater exploitation for karst-suffosive process development. Moscow. *Nauka*. 133 p.

Index

D

E

F

G

H

I

K

L

M

N

O

P

R

S

T

U

V

W

Z